Intelligent Systems Reference Library

Volume 194

The aim of this series is to publish a Reference Library, including novel advances and developments in all aspects of Intelligent Systems in an easily accessible and well structured form. The series includes reference works, handbooks, compendia, textbooks, well-structured monographs, dictionaries, and encyclopedias. It contains well integrated knowledge and current information in the field of Intelligent Systems. The series covers the theory, applications, and design methods of Intelligent Systems. Virtually all disciplines such as engineering, computer science, avionics, business, e-commerce, environment, healthcare, physics and life science are included. The list of topics spans all the areas of modern intelligent systems such as: Ambient intelligence, Computational intelligence, Social intelligence, Computational neuroscience, Artificial life, Virtual society, Cognitive systems, DNA and immunity-based systems, e-Learning and teaching, Human-centred computing and Machine ethics, Intelligent control, Intelligent data analysis, Knowledge-based paradigms, Knowledge management, Intelligent agents, Intelligent decision making, Intelligent network security, Interactive entertainment, Learning paradigms, Recommender systems, Robotics and Mechatronics including human-machine teaming, Self-organizing and adaptive systems, Soft computing including Neural systems, Fuzzy systems, Evolutionary computing and the Fusion of these paradigms, Perception and Vision, Web intelligence and Multimedia.

** Indexing: The books of this series are submitted to ISI Web of Science, SCOPUS, DBLP and Springerlink.

More information about this series at http://www.springer.com/series/8578

Abhaya Pal Singh · Dipankar Deb ·
Himanshu Agrawal · Valentina E. Balas

Fractional Modeling and Controller Design of Robotic Manipulators

With Hardware Validation

Springer

Abhaya Pal Singh
Symbiosis International
(Deemed University), Symbiosis
Institute of Technology
Pune, India

Dipankar Deb
Electrical Engineering Department
Institute of Infrastructure Technology
Research and Management (IITRAM)
Ahmedabad, Gujarat, India

Himanshu Agrawal
School of Electrical Engineering
Computing and Mathematical
Sciences, Curtin University
Perth, Australia

Valentina E. Balas
"Aurel Vlaicu" University of Arad
Arad, Romania

ISSN 1868-4394 ISSN 1868-4408 (electronic)
Intelligent Systems Reference Library
ISBN 978-3-030-58246-3 ISBN 978-3-030-58247-0 (eBook)
https://doi.org/10.1007/978-3-030-58247-0

This Springer imprint is published by the registered company Springer Nature Switzerland AG
The registered company address is: Gewerbestrasse 11, 6330 Cham, Switzerland

Dedicated to my parents Mahendra Pal Singh and Nirmala Singh, wife Pallavi Srivastava, sisters Archana, Anuradha and my loving nephew Adway for their encouragement, care, love, and support.

Abhaya Pal Singh

Dedicated to my wife Indulekha and my son Rishabh for providing me unending love and support.

Dipankar Deb

Dedicated to my wife Nidhi and my daughters Yashi and Aditi for their unconditional love and support.

Himanshu Agrawal

Dedicated to my husband Marius for patience, love, friendship, and humor.

Valentina Emilia Balas

Preface

Robotic Manipulators are equipped for performing any task with high accuracy and exactness compared to manual operators. As the demand for more precise manipulators is increasing, the main objective of this monograph is to introduce about improvement in the performance of Robotic Manipulators and their Control. The traditional mathematical model of any system is derived by using Integer Order (IO) Calculus, but it limits the performance of the system as every system cannot be represented only by using integer numbers. In 1695, L'Hopital stated that a more accurate model of a system could be derived using Fractional Order (FO) Calculus. In 1994, Podlubny introduced the first Fractional Order PID Controller (FOPID) designed for Fractional Model of a system. The trail of the research on Fractional Calculus keeps growing with contributions from Sonin, Krug, Abel, Fourier, Lacroix, Grunwald, Leibniz and Letnikov till the nineteenth century and this was the basis of fractional calculus.

It is now established Fractional Order systems can depict the dynamical behavior of systems, processes, and notions over an extensive range of frequency and time with the very computable and concise model. Hence, the model of a Robotic Manipulator is derived using Fractional Order Calculus to improve its performance. Similarly, Fractional Order (FO) PID Controllers are also designed for different Integer Order (IO) and Fractional Order Robotic Manipulators. Various other approaches are also used to design the controller for the FO model of a given Robotic Manipulator. The performance of the IO model and FO model of a given system is compared to identify a better model. Later, an algorithm is proposed to identify the parameters for the best fractional model of any given system. Finally, the experimental validation of the outcome is provided to support the simulation result.

The first objective of this monograph is to give exposure to the feasibility of fractional order modeling of different robotic systems. This is achieved by deriving Fractional Order Models for various Robotic Manipulators, and their performances are compared with their respective Integer Order Models. The FO model is derived using the definitions of the Laplace transformation of FO functions. Furthermore, these FO model is approximated to the IO model to give the physical significance of the considered system, and this is discussed in Chap. 2.

The other two objectives of this monograph are to explore the feasibility of Fractional Order Controller design for IO and FO Model of different robotic systems. Different types of Controllers are designed for various sets of robotic manipulators and is discussed in Chap. 3 (talks about PID and Fractional PID controllers) and Chap. 4 (talks about Fractional Model Predictive Controller (FMPC)).

A Pole Placement Method of Controller Design is discussed for FO Model of Single Flexible Link Robotic Manipulator, and the outcome is discussed in Chap. 5 and performance is analysed in time and frequency domain both.

An algorithm is proposed to select the best FO model corresponding to a system in Chap. 6. The performance merit of this algorithm is examined on different existing systems, and it is observed that the model obtained by this algorithm improves the performance by more than 80% in terms of settling time and overshoot. The FO model validation is also explored in Chap. 6. The fractional order model of 2D Serial Flexible Link robotic manipulator is derived using the algorithm proposed in Chap. 6. The FO controller is then designed for the FO model of 2DSFL to control the angular positions of the two links. Simulation and experimental results are compared and found that the FO controller design for the FO model of this robotic manipulator is precisely tracking the desired angular position in minimum time with minimum overshoot.

In Chap. 7, a basic idea of how to design adaptive fractional controller for robotic systems have been discussed. A Model Adaptive Reference Control (MARC) strategy is opted to find the control law. Pendulum on a cart system, 2D gantry crane system, Missile launching vehicle, single link robotic manipulator and 2 DOF serial link robotic manipulator have been considered to design the adaptive fractional controller design.

The outcomes obtained in monograph provides compelling evidence that Fractional Order Modeling derived by using the proposed algorithm improves the performance of any system by more than 80% in terms of settling time. Furthermore, it is experimentally validated that FOPID Controller designed for the FO Robotic Manipulator is precisely tracking the desired angular position in minimum time with minimum overshoot.

We would like to thank Dr. Thomas Ditzinger, Editorial Director of Interdisciplinary and Applied Sciences Engineering and Prof. Dr. Janusz Kacprzyk and Dr. Lakhmi C. Jain, Series Editors of Intelligent Systems Reference Library for acceptance of the publication of this book with Springer. We also want to thank the production team of Springer, in particular, the project coordinator Ms. Varsha Prabakaran for her professional and kind assistance.

Thanks and appreciations also go to Dr. Ketan Kotecha, Dean and Director, Symbiosis Institute of Technology, Pune and Dr. N. N. Bhuptani, Registrar, IITRAM, Ahmedabad for providing necessary infrastructure and resources to accomplish the research work.

Chapter 1
Introduction

Fractional calculus is a variant of differential calculus wherein integrals and derivatives are in fractional order, and can be implemented for dynamic systems modeled both in integer order and fractional order equations. Fractional Calculus (FC) have generated considerable recent research interest and has found various applications especially in control systems. Fractional order modeling of a system results in better performance compared to its Integer model. Integer order model is itself an approximation of some FO equivalent model. Implementation of Fractional model is comparatively complex, but results in improved performance like robustness, transient stability, noise filtering and disturbance rejection. Designing controller for a given system is another major issue, and fractional order controllers have been demonstrated commendable performance even in this field. Fractional order PID (FOPID) controllers were first mentioned in Podlubny's work [1, 2] in which a fractional-order integrator and a fractional-order differentiator is designed for a fractional-order system. FOPID controllers generally have more tunable parameters and adjust the controller based on the system requirements, thus providing more space to implement the controller precisely and more accurately to satisfy system requirements.

Renowned mathematicians like Liouville, Weyl, and Riemann have significantly contributed to FC. The trail of the research on FC has grown over the centuries with contributions from Sonin, Krug, Abel, Fourier, Lacroix, Grunwald, Leibniz and Letnikov till the 19th century which formed the basis of FC [3]. The first monograph of FC was published only in 1974 [4]. It is now understood that FO systems can depict dynamical behavior of systems, processes, and notions over an extensive range of frequency and time with computable and concise model [5, 6]. Advanced control systems handle undesired and unsuitable characteristics of the plant response using FO controllers [7, 8], and so do the work on signal filtering methods [9].

A FC approach to Newtonian mechanics is developed which generalizes the Newtonian mechanics and the Fractional Langevin equation as non-local models [10]. A FC approach is also applied in differential and integral vector operations [11]. A

A. P. Singh et al., *Fractional Modeling and Controller Design of Robotic Manipulators*, Intelligent Systems Reference Library 194,
https://doi.org/10.1007/978-3-030-58247-0_1

fractional generalization of differential calculus of differential forms is discussed. A FC approach in field theory is discussed, which generalizes the Maxwell's equations and uses fractal differential equations [12].

A possible generalization of Newton's second law of motion using FC is considered and applied to a body subjected to a constant force. The authors also reported the second application of FC to Newtonian gravity which is a generalized fractional gravitational potential and might be used as a tool to model galactic rotation curves for dark matter problem. A conformable fractional derivative and integral is considered to discuss the fractional Newtonian mechanics [13]. Some mechanical problems such as fractional harmonic oscillator problem, the fractional damped oscillator problem and the forced oscillator problem are discussed in one-dimensional fractional dynamics, and a possible generalization in fractional space is reported [13]. The concept of FC in the context of static electric potential is studied and a new FO approximation is developed [14]. A special class of fractional derivatives is considered for motion under gravity with and without resistance to arrive at a new notion of time that depends on the fractional parameter [15]. Such results in the literature show the limitations and use of generalized derivatives and the corresponding fractional derivatives. Fractional order explains the system's behavior more clearly than integer calculus and fractional calculus is the generalization of integer calculus.

1.1 Applications of Fractional Calculus

Fractional calculus is applicable in diverse applications; in almost all the field of science and engineering as illustrated in Fig. 1.1.

For instance, FC has shown tremendous potential in improving various operations in the medical field. Wang et al. have applied a novel FO differentiation model to CT image processing for post-processing purposes using a weighted combination of the fractional order Perona-Malik model and the fractional-order total variation model to obtain a better performance over other existing methods through simulation and clinical data experiments [16]. Li et al. present a new medical image enhancement method that adjusts the FO according to the dynamic gradient feature of the entire image to extract the edges accurately and enhance them while preserving smooth areas and weak textures, thereby enabling better diagnosis [17]. Different applications of FC in the medical and related fields, is illustrated in Fig. 1.2.

Certain characteristics of FC that make it suited for bio-engineering and development of Fractional-order linear and nonlinear models for bio-engineering applications along with an effective method for numerical solution is discussed [18]. Some applications of FO calculus in biomedical signal processing with emphasis on the ability of a mathematical tool to remove noise, and generate fractal signals has also been studied [19].

An improved Variable Order Fractional Centered Difference (VOFCD) scheme has been derived with a dynamically adjustable fractional differential order based on second order Riesz fractional differential operator and a Lagrange 3-point inter-

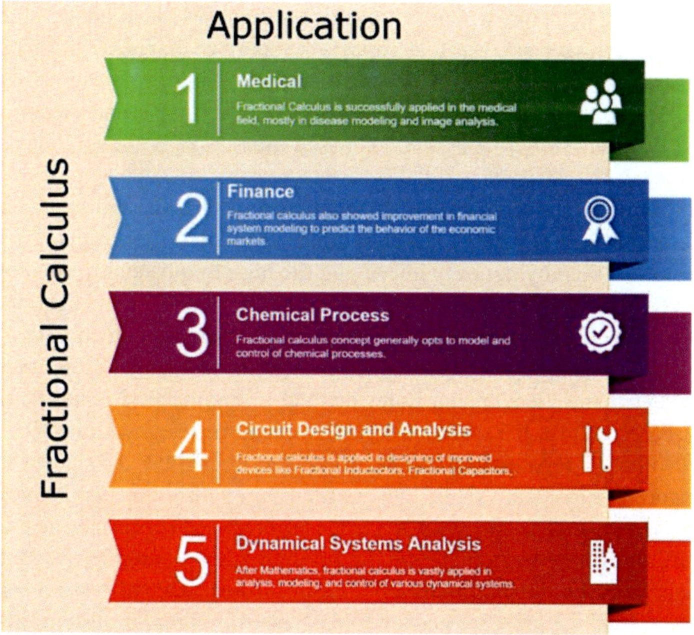

Fig. 1.1 Fractional calculus applications

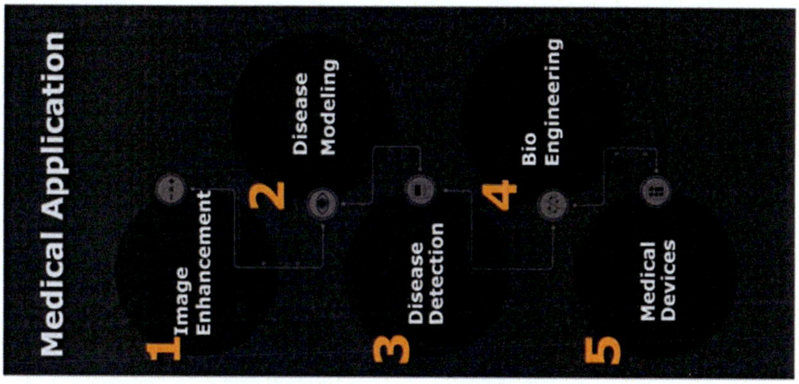

Fig. 1.2 Fractional calculus applications in medical field

polation formula for enhancement of medical images related to patients with stroke and Parkinson's disease. The experiments show that the proposed method improves fractional differential mask with a higher signal to noise ratio value than the other fractional differential mask operators [20]. A fractional differential covering template applied to low-dimension medical image processing is presented to enhance the image through the basic definition of the fractional calculus. The experiments show that the method is simple, can enhance the medical images efficiently unlike the traditional edge detectors which are sensitive to noise. Medical image edge detection is a key step of image analysis that impacts the speed of diagnosis and accuracy [21]. A comparative study of four fractional order filters for edge detection has been also studied. The mean square error (MSE) of the detected images and execution time are adopted as criteria for comparison. FC based filter mask is applied on each pixel of noisy image [22].

Another method considers the surrounding information and structural features of different pixels, as well as the directional derivative of each pixel in constructing the masks, thereby not only improving the high frequency information, but also the low frequency image information [23]. A fractional-order infection model of $HIV-1$ infection of $CD4^+$ T cells is applied in disease detection to analyze the effect of changing the average number of viral particles N with different sets of initial conditions using a Generalized Euler method (GEM) to find a numerical solution [24]. A FO model for the spread of human immunodeficiency virus (HIV) infection and the effect of screening of unaware infected individuals, is studied using local asymptotic stability analysis of the disease-free equilibrium [25]. A conventional lumped element circuit model of electrodes is analyzed for further investigation by generalization of the fractional order of differentiation for an improved bioelectrode behaviour, but recent experimental studies of cardiac tissue also suggests that additional mathematical tools are needed to describe this complex system [26].

Fractional calculus is also efficient in modeling of financial systems. Different applications of FC in the financial systems, is illustrated in Fig. 1.3.

The issues of synchronization and anti-synchronization for fractional chaotic financial system with market confidence through active control approach, is discussed [27]. A fractional incommensurate order financial system with interesting dynamic behaviors like chaotic motions, periodic motions and fixed points are discussed [28–30], and the chaotic behavior are eliminated using nonlinear feedback

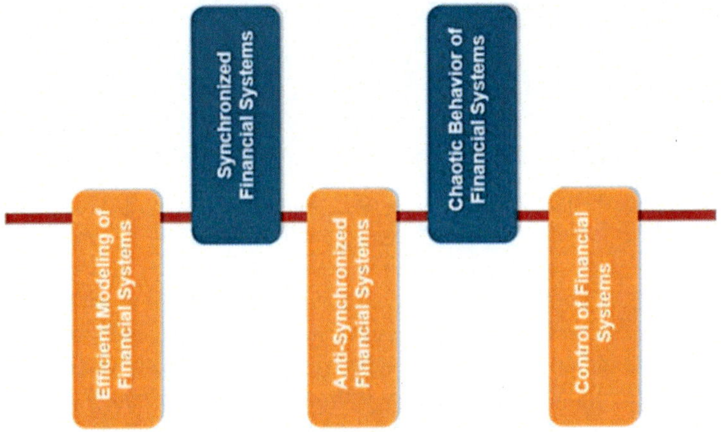

Fig. 1.3 Fractional calculus applications in financial systems

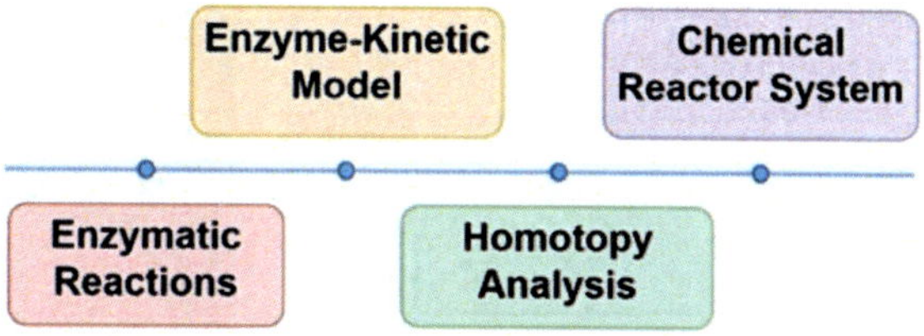

Fig. 1.4 Fractional calculus applications in process control

control and Lyapunov linearization method. Financial systems are known to have irregular and erratic fluctuations due to diverse influences and often result in economic crisis and huge financial losses. A FO fuzzy control policy is employed to suppress the chaotic dynamics of a representative FO financial system. An intelligent Regrouping Particle Swarm Optimization (Reg-PSO) is used to design the numeric weights of the control policy and the methodology is demonstrated by simulations [31]. A sliding mode controller (SMC) for a FO chaotic financial system is considered [32] wherein a sliding surface is determined and a sliding mode control law is derived to make the FO financial system asymptotically stable.

A new fraction-integer integral switching surface is constructed to facilitate stability analysis of the closed-loop system. A novel integer-order chaotic financial system is proposed by considering market confidence into a three-dimensional financial system [33]. A FO approach to model the growth of national economies (of Spain and Portugal), namely, their gross domestic products (GDPs) over the last five decades is studied [34]. A new approach to macroeconomic modeling is presented to model the behavior of national economies of United Kingdom, Canada, and Australia using (i) state-space modeling adopted from the classical control theory wherein the gross domestic product (GDP), inflation, and unemployment rate (UE) were chosen as the state variables, (ii) fractional-order differential equations to fit the available economic data, and (iii) orthogonal distance fitting method to identify the system parameters [35]. Another major filed of application of FC has been in chemical processes. Different applications of FC in the process control, are illustrated in Fig. 1.4.

Enzymes act as catalysts and are significant in controlling the characteristics of chemical and biochemical reactions. In [36], a dynamical Brusselator reaction-diffusion system arising in triple collision and enzymatic reactions using qhomotopy analysis transform method, and time fractional derivative has been studied. In [37], an enzyme-kinetic mathematical model is formulated using FO derivatives and Optimal control mechanism is incorporated into to maximize the product output. In [38], a generalized fractional cubic isothermal auto-catalytic chemical system with new variants of fractional time derivatives using Homotopy Analysis Transform are reported. The convergence of Homotopy Analysis Transform Method is also studied by computing the residual error function. An alternative tool based on FO differential equations for biochemical reactor identification using previously reported experimental data, is reported [39]. The key aim is to analyze a new fractional model of chemical kinetics related to the newly developed FO Atangana-Baleanu derivative having nonsingular and non-local kernel, and to derive a numerical solution [40].

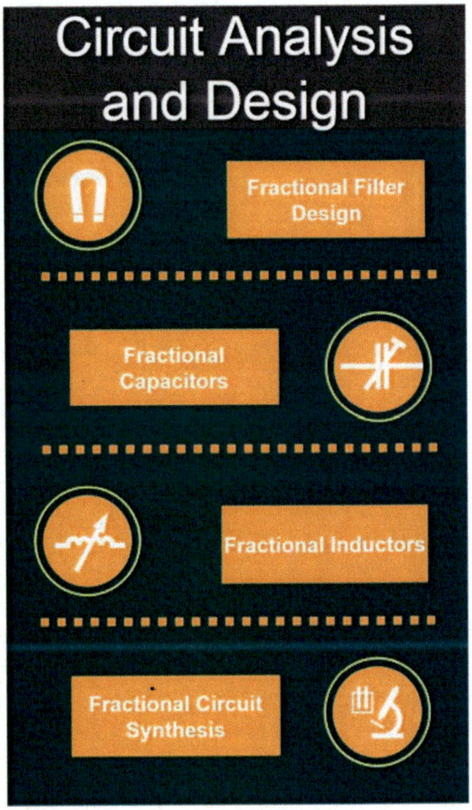

Fig. 1.5 Fractional calculus application in circuit analysis and design

In [41], the dynamic simulation and optimization of chemical processing systems modeled in terms of FO differential equations is addressed. A new numerical method based on Gaussian quadrature is proposed to address larger-scale fractional differential equations. In [42], chaos control, stability analysis and function projective synchronization between a chaotic chemical reactor system and a FO chaotic chemical reactor system with uncertain parameters are performed with the help of the Caputo derivative and Lyapunov stability analysis. Nonlinear control method is used to achieve function projective synchronization and chaos control of considered chaotic chemical reactor system. Fractional calculus is applied to many real-time circuits to improve their performances by introducing a new fractional inductors and fractional capacitors, as illustrated in Fig. 1.5.

A novel configuration of FO filter topologies realized through the concept of companding filtering is introduced, and the performance of the FO filters is evaluated through simulation [43]. In [44], a new infinite impulse response (IIR)-type digital fractional order differentiator (DFOD) is proposed by using a family of first-order digital differentiators expressed in the second-order IIR filter form. In [45], the design of FO Simpson digital integrator is investigated. The conventional transfer function of second-order IIR Simpson integrator is factorized into product of first-order factors, and then each factor of product is fractionalized by taking fractional power and binomial series expansion. The use of a compact integer-order transfer function approximation of the FO Laplacian operator s^{α} to realize fractional-step filters is possible using FC approach [46]. In [47], the use of nonlinear least squares optimization to approximate pass-band ripple characteristics of traditional Chebyshev low-pass filters with FO steps in the stop-band is studied. FO energy storage elements are considered to improve the circuit performances [48]. A general procedure to obtain Butterworth filter specifications in the FO domain where an infinite number of relationships are possible due to the additional independent FO parameters which increase the filter degrees-of-freedom [49].

An electrolytic process in the perspective of FO capacitors is reported [50] with several experiments for measuring the electrical impedance of the devices. The results analyzed through frequency response, reveal capacitances of FO that can constitute an alternative to the classical integer order elements. The experimental results from a fabricated integrated circuit of FO capacitor emulators are reported to be successful [51]. The chip contains emulators of capacitors of orders 0.3, 0.4, 0.5, 0.6 and 0.7 with nF pseudo-capacitances that can be adjusted through a bias current. A survey on FO electric circuit models that best fit experimentally collected impedance data from energy storage and generation elements including super-capacitors, batteries, and fuel cells, finds that the performance of these devices improve with the use of fractional element [52]. The hardware realization and performance study of fractional inductors of order $0 < \alpha < 2$ have indicated positive outcomes for circuit applications [53]. In [54], it is found that a FO model with only seven parameters can accurately describe the responses of networks composed of more than 70 nodes and 200 branches with 100 resistors and 100 capacitors. Generalization of conventional RC and RL circuits in FO sense to improve their performances, is reported [55]. A new FO parallel resonator circuit whose resonating frequency is tuned by the change in the coefficient of a FO element, perform better [56]. The overall response of any dynamic system rely on the system model accuracy, and it worsens if an inaccurate system model is chosen. Applications of FC on dynamical systems can be categorized as illustrated in Fig. 1.6.

Many real-time dynamical systems possess FO behavior. In [57], the modeling of the material aging within the limitation of FC is discussed. A method of capacitor characteristic control using FO models for electrode behavior description is studied by opting for FC strategy [58]. Fractional modeling of love and happiness is also derived in history [59, 60]. Fractional Order Modeling of a system can provide a better response as it can attain more accurate values of the required parameters [61–63]. A FO system is considered to design an accurate controller. It is better to represent the system model into FO for a more precise system response [64].

The fractional integro-differential, that is, FC operators are intriguing in robotics and control engineering. The transient response and frequency of the FO integral model and its application in control engineering are presented [65]. Vinagre et al. [66] proposed a generalized PID controller represented as $PI^{\lambda}D^{\mu}$ controller which involves a differentiator of the order of 'm' and an integrator of the order of 'l'.

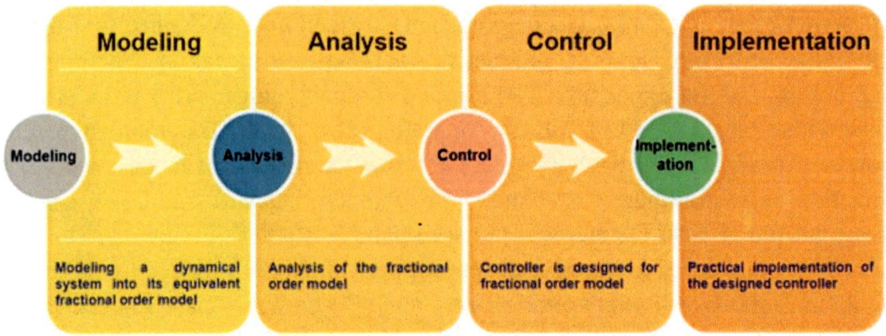

Fig. 1.6 Applications of fractional calculus on dynamical systems

Some tuning techniques of FO controller is also reported [67] which give a more adaptable tuning strategy. Recently, FO controllers have been used in control of electric vehicles [68], magnetic levitation [69], DC motor performance [70], and to control the dynamic behavior of power systems [71]. In the process industries, Model Predictive Control (MPC) method is generally utilized. Deng et al. considered the design of a generalized predictive controller for a fuel cell using FC [72]. Others have focused on the design of an MPC for FO discrete-time systems [73, 74]. Rhouma et al. designed MPC for an existing model of a thermal system [75]. This book utilizes a similar methodology to design Model Predictive Controller for Robotic Manipulators. A novel fractional model of robotic manipulators is derived in this work, and then an appropriate FO controller is developed for this model.

1.2 Studies on Robotic Manipulators

An industrial robot comprises of different parts such as robot manipulator, power supply, and controllers. Robotic manipulators consists of two sections: (i) Robot Arm and (ii) Body which together move and position parts within a work envelope. Robot manipulators are created from a sequence of link and joint combinations, and therefore in this book we experimentally demonstrate performance of both flexible link and flexible joint manipulators. The links are the rigid members connecting the joints, or axes. The axes are movable components of the robotic manipulator to provide relative motion between adjoining links. Significant control problems appear in robotic manipulators like vibration, and static deflection from external effects, and designing errors which can decrease the accuracy, increase settling time, and complicate the controller design scheme. Robotic manipulators have been a point of research since many years to do some work which may be related to spray painting, pick and place an object, bomb diffusion, moving from an hazardous environment, or even in the kitchen for chopping. Hence, control of such robotic manipulators are important in performing the desired task efficiently. The following subsections will give a brief idea of FO modeling and controller design for robotic manipulators.

1.2.1 Supported by Simulations

Control of robotic manipulators using FC has been looked at since the early 90's when Oustaloup et al. [76] introduced the concept of CRONE control which means 'Non Integer Order Robust Control' using a mathematical principle, namely, non integer derivation applied to robot control. In 1995, Machado [77] came out with a motion control technique using FO PID control which is applicable to robotic manipulators. Subsequently, again Machado et al. [78] developed a position and force control of manipulators using FC. A FO hybrid control of robot manipulators proposed by Machado et al. in 1998, uses fractional derivative and integral (FDI) control strategy applied on an integer model, and is supported by simulation results.

Later in 2001, Duarte et al. [79] proposed a FO trajectory control of robotic manipulators. In 2001, Melchior et al. [80] proposed the CRONE control toolbox for fractional path planning of robotic manipulators. Duarte and Machado [81] observed chaotic phenomena using FC in trajectory control of manipulators, and later proposed a pseudoinverse trajectory control of manipulators using FC [82]. Orsoni et al. [83] also come-up with a new approach of path tracking using FC. Subsequently, many researchers have come-up with different techniques of fractional modeling and fractional control. Ferreira et al. [84] proposed a strategy of hybrid control of robotic manipulators using FC. Pires et al. [85] investigated the FO dynamics of planner robotic manipulators. Silva et al. [86] achieved improved performance with a fractional controller designed for hexapod robots. Valerio et al. [87] proposed a genetic algorithm based fractional controller design strategy for two flexible link robotic manipulators. Lazarevic [88] has talked about stability procedure for robotic manipulators using FC. Feliu et al. also proposed a FO control design strategy for flexible link robotic manipulators which provides improved performance compared to traditional controllers [89]. Efe, in 2008, proposed a novel parameter adjustment strategy to improve performance of sliding mode control using FC applied to robot arm [90]. Silva et al. presented a FO controller design method for joint control of hexapod robot [91]. Lima et al. depicted fractional dynamical behavior of mechanical manipulator [92]. Monje et al. came up with a tuning rule for fractional controllers applied to robotic manipulators [93]. Marcos et al. proposed an idea for fractional dynamics in trajectory control of manipulators [94]. Other applications of FC in manipulators path planning and control have been proposed [95, 96].

Delavari et al. proposed a novel controller for robotic manipulator based on FC [97]. Dumlu et al. proposed a trajectory tracking control of 3 DOF robotic manipulator [98]. Sharma et al. found that FO fuzzy controller designed by cuckoo search algorithm is superior [99]. Mujumdar et al. [100] proposed an approach of FO sliding mode control design for fractional model of robotic manipulators using MATLAB Simulink. Sharma et al. have shown the performance analysis of a robotic manipulator with payloads, using FC [101]. Viola and Angel proposed robustness analysis, identification and control of robotic manipulators [102]. An optimized fractional PID is designed for single link manipulator [103]. A trajectory tracking control of robotic manipulator is proposed using fuzzy and FC [104]. Two layered fractional fuzzy controller is also designed [105]. Various fractional controller have been designed using PID [106–118], terminal sliding mode control [119, 120], adaptive control [121], fuzzy control [122, 123], and MPC control [124, 125] for robotic manipulators.

1.2.2 Supported by Experiments

Monje et al. [126] proposed a tip position control of a lightweight flexible manipulator using a FO controller, and used the methodology of FO PD controller to control single flexible link manipulator. In this study, a fractional controller is designed for integer model and is supported by experimental results. In 2010, Barbosa et al. proposed an effect of fractional orders in the velocity control of a servo system. In this work, a FO PID controller (with different combination) is used to control integer model servo motor system, and is supported by experimental results. Luo et al. in 2011 proposed an experimental study of FO PD controller synthesis to control well defined fractional models of Membrane Charging and supported by experimental results. An adaptive FO controller is designed by Nikdel et al. [127], for integer model of robotic manipulator and is validated by experiments. Wang et al. [128] designed a fractional controller using continuous FO sliding mode control which is supported by an experiment. An observer based sliding mode controller is designed for robotic manipulators by Mujumdar et al. [129] and the fractional model is validated experimentally. Terminal sliding model controller design strategy is explained in [130] and is supported by experiments. Some more practical implementation of FO controllers for integer model of robotic manipulators are available [131, 132].

1.3 Objectives and Scope

This scope of the book is focused on the FO modeling and controller design for robotic manipulators. For different robotic systems, the objectives are

(i) To explore the feasibility of FO modeling,
(ii) To explore the feasibility of FO Controller design for Integer model,
(iii) To explore the feasibility of FO Controller design for FO model,
(iv) Experimental validation of the results.

The fractional model proposed in this book is aimed at improving the overall system performance, as verified by simulations and some experiments. Various robotic manipulators such as Inverted Pendulum system, Double Inverted Pendulum system, 2D Gantry Crane system, Missile Launching Pad system, 2 DOF Serial Link Robotic manipulator, and 2 DOF Serial Joint Robotic manipulator are considered for modeling and controller design. The results are evaluated experimentally on 2 DOF Serial Link Robotic manipulator, and 2 DOF Serial Joint Robotic manipulator. Albeit, the scope of this book is limited to robotic manipulators, the study can also be extended to various other systems like process control, signal processing, image processing, etc. Studies on Robotic manipulators performed in Sect. 1.2 reveal:

- The research work for modeling of robotic manipulators is largely focused on IO Calculus. Inconsiderable amount of research work has been carried out in the area of Fractional Modeling of Robotic Manipulators.
- There is literature available on designing the IO and FO Controllers for an Integer Model of robotic manipulators, but design of an Integer/Fractional Order Controller for the FO Model of robotic manipulators has not been reported yet.
- Experimental validation of the results expected for robotic manipulators is scarce.

Therefore, it is required to derive the FO model for Robotic Manipulators for precise system response. It is also required to design the Fractional Controller and Integer Controller for the Fractional model of Robotic Manipulators for improved overall system response. Experimental validation of the results is needed. We address many specific questions related to robotic manipulators as relevant to FC:

(i) How to derive a FO model for Robotic Manipulators?
(ii) How to design a FO Controller for the FO Robotic Manipulators?
(iii) What approach should be opted for experimental validation of the results?

The major contributions of this book include the introduction of a novel fractional model of robotic manipulators and various control design strategies for those models. Integer modeling of the dynamical behavior of robotic systems, is presented along with fractional modeling of the dynamical behavior. We also present a PID and fractional PID controller design for robotic systems. FMPC controller strategy design for fractional model of robotic systems, followed by validation of robustness of such a controller, is also presented. An algorithm proposed to obtain the best FO model of a system to design the Controller, is validated in the book with various examples. A pole placement method of controller design for fractional model of robotic manipulators is also developed. The efficacy of the proposed algorithm is validated experimentally in different ways:

(i) FO controller is designed for IO Model of 2 DOF Serial link (2DSFL) robotic manipulator and 2 DOF Serial Joint (2DSFJ) robotic manipulator,
(ii) FO model derivations of the 2DSFL and 2DFFJ robotic manipulators using the proposed algorithm and its controller design, are validated, and
(iii) FO controller design for the FO model of a 2DSFL robotic manipulator.

This book introduces a novel method of FO modeling of a Robotic Manipulator and compares response with Integer Model. Furthermore, Fractional Controller and Integer Controller are designed for the Fractional model of Robotic Manipulators, and an algorithm is proposed to obtain the best FO model of a system to design the Controller. The rest of the book is organized as follows:

Chapter 2 presents mathematical modeling of benchmark systems like POAC system and DIPOAC system. The equations of motion are obtained using Euler-Lagrange formulation. This chapter also introduces modeling of robotic manipulators like 2-D gantry crane systems and Missile launching vehicle/pad. These two systems are modeled, and dynamic modeling equations are obtained by using Euler-Lagrange formulation. It also gives the concepts of FO modeling of different robotic systems like POAC, 2D Gantry Crane and MLV. Equivalent fractional dynamical equations are obtained using the definition of fractional Laplace Transform. Oustaloup recursive approximation is used to find the equivalent transfer function. Chapter 3 provides the design steps of FO PID and PID controllers for robotic manipulators. The controllers are designed for integer order model of robotic manipulators. From the simulation experiment it found that fractional PID performs better compared to traditional integer order PID for fast system response.

Chapter 4 discusses the design of FMPC for robotic manipulators and an FMPC controller is proposed to control the dynamical behavior of POAC, 2D Gantry Crane system, and MLV system. The controller is designed for fractional equivalent model of the robotic manipulators. The usage of FMPC controller to control dynamic behavior of robotic manipulators is novel. The proposed FMPC controller provides better performance and is robust to the system parameter variations compared to traditional MPC. This work is followed by a presentation of Pole-Placement control technique for FO model of a robotic system in Chap. 5. A fractional model of single link robotic manipulator is derived and the system response is analyzed. It is found that a pole placement method of controller design can also be achieved for fractional model of robotic manipulators.

A novel algorithm is proposed in Chap. 6 to find the best possible FO system model. This algorithm is applied to existing FO systems and corresponding FO model is found to improve the system performance by more than 80% compared to existing fractional models. A fractional equivalent model of mass spring system and DC motor control system is also presented, followed by experimental validation on two different robotic manipulators, i.e., 2DSFL and 2DSFJ robotic manipulators. The experimental outcomes support the simulation results and validate that FO controller designed for fractional equivalent model of the robotic manipulator, gives superior response as compared to other controller strategies. Chapter 7 gives the basic idea of how to design adaptive fractional controller for robotic systems. In this chapter Model Adaptive Reference Control strategy is opted to find the control law. Pendulum on a cart system, 2D gantry crane system, Missile launching vehicle, single link robotic manipulator and 2 DOF serial link robotic manipulator have been considered to design the adaptive fractional controller design.

References

1. Podlubny, I.: Fractional-order systems and $PI^{\lambda}D^{\mu}$-controllers. IEEE Trans. Autom. Control **44**(1), 208–214 (1999)
2. Podlubny, I.: Fractional-order systems and fractional-order controllers. Inst. Exp. Phys. Slovak Acad. Sci. Kosice **12**(3), 1–18 (1994)
3. Grunwald, A.K.: Uber "begrenzte" Derivationen und deren Anwendung. Zeitschrift för Mathematik und Physik **12**, 441–480 (1867)
4. Oldham, K., Spanier, J.: The fractional calculus theory & applications of differentiation and integration to arbitrary order, vol. 111. Elsevier (1974)
5. Ortigueira, M.D.: An introduction to the fractional continuous-time linear systems: the 21st century systems. IEEE Circuits Syst. Mag. **8**(3), 19–26 (2008)
6. Fenander, A.: Modal synthesis when modeling damping by use of fractional derivatives. AIAA J. **34**(5), 1051–1058 (1996)
7. Bohannan, G.W.: Analog fractional order controller in a temperature control application. IFAC Proc. Vol. **39**(11), 40–45 (2006)
8. Machado, J.A.: Discrete-time fractional-order controllers. Fract. Calc. Appl. Anal. **4**, 47–66 (2001)
9. Ortigueira, M.D., Serralheiro, A.J.: A new least-squares approach to differintegration modeling. Signal Process. **86**(10), 2582–2591 (2006)
10. Baleanu, D., Golmankhaneh, A.K., Nigmatullin, R., Golmankhaneh, A.K.: Fractional newtonian mechanics. Central Eur. J. Phys. **8**(1), 120 (2010)
11. Tarasov, V.E.: Fractional vector calculus and fractional Maxwell's equations. Ann. Phys. **323**(11), 2756–2778 (2008)
12. Golmankhaneh, A.K., Baleanu, D.: Heat and Maxwell's equations on Cantor cubes. Rom. Rep. Phys. **69**, 1–11 (2017)
13. Chung, W.S.: Fractional Newton mechanics with conformable fractional derivative. J. Comput. Appl. Math. **290**, 150–158 (2015)
14. McBride, A.: Advances in fractional calculus: theoretical developments and applications in physics and engineering (2008)
15. Mingarelli, A.B.: On generalized and fractional derivatives and their applications to classical mechanics. J. Phys. A: Math. Theor. **51**(36), 365204 (2018)
16. Wang, Y., Shao, Y., Gui, Z., Zhang, Q., Yao, L., Liu, Y.: A novel fractional-order differentiation model for low-dose CT image processing. IEEE Access **4**, 8487–8499 (2016)
17. Li, B., Xie, W.: Adaptive fractional differential approach and its application to medical image enhancement. Comput. Electr. Eng. **45**, 324–335 (2015)
18. Petráš, I.: An effective numerical method and its utilization to solution of fractional models used in bioengineering applications. Adv. Differ. Equ. **2011**(1), 652789 (2011)
19. Ferdi, Y.: Some applications of fractional order calculus to design digital filters for biomedical signal processing. J. Mech. Med. Biol. **12**(02), 1240008 (2012)
20. Yu, Q., Vegh, V., Liu, F., Turner, I.: A variable order fractional differential-based texture enhancement algorithm with application in medical imaging. PloS one **10**(7), e0132952 (2015)
21. Zhang, X.X., Lu, Y.: Medical image edge detection method based on fractional differential coefficient [J]. J. Dalian Jiaotong Univ. **6** (2009)
22. Saadia, A., Rashdi, A.: Incorporating fractional calculus in echo-cardiographic image denoising. Comput. Electr. Eng. **67**, 134–144 (2018)
23. Guan, J., Ou, J., Lai, Z., Lai, Y.: Medical image enhancement method based on the fractional order derivative and the directional derivative. Int. J. Pattern Recognit. Artif. Intell. **32**(03), 1857001 (2018)
24. Arafa, A.A.M., Rida, S.Z., Khalil, M.: Fractional modeling dynamics of HIV and CD4+ T-cells during primary infection. Nonlinear Biomed. Phys. **6**(1), 1 (2012)
25. Babaei, A., Jafari, H., Ahmadi, M.: A fractional order HIV/AIDS model based on the effect of screening of unaware infectives. Math. Methods Appl. Sci. **42**(7), 2334–2343 (2019)

26. Magin, R.L., Ovadia, M.: Modeling the cardiac tissue electrode interface using fractional calculus. J. Vib. Control **14**(9–10), 1431–1442 (2008)
27. Huang, C., Cao, J.: Active control strategy for synchronization and anti-synchronization of a fractional chaotic financial system. Phys. A: Stat. Mech. Appl. **473**, 262–275 (2017)
28. Wang, Z., Huang, X., Shi, G.: Analysis of nonlinear dynamics and chaos in a fractional order financial system with time delay. Comput. Math. Appl. **62**(3), 1531–1539 (2011)
29. Hajipour, A., Tavakoli, H.: Analysis and circuit simulation of a novel nonlinear fractional incommensurate order financial system. Optik **127**(22), 10643–10652 (2016)
30. Abd-Elouahab, M.S., Hamri, N.E., Wang, J.: Chaos control of a fractional-order financial system. In: Mathematical Problems in Engineering (2010)
31. Pan, I., Korre, A., Das, S., Durucan, S.: Chaos suppression in a fractional order financial system using intelligent regrouping PSO based fractional fuzzy control policy in the presence of fractional Gaussian noise. Nonlinear Dyn. **70**(4), 2445–2461 (2012)
32. Dadras, S., Momeni, H.R.: Control of a fractional-order economical system via sliding mode. Phys. A: Stat. Mech. Appl. **389**(12), 2434–2442 (2010)
33. Wang, Z., Huang, X., Shen, H.: Control of an uncertain fractional order economic system via adaptive sliding mode. Neurocomputing **83**, 83–88 (2012)
34. Tejado, I., Valério, D., Pérez, E., Valério, N.: Fractional calculus in economic growth modelling: the Spanish and Portuguese cases. Int. J. Dyn. Control **5**(1), 208–222 (2017)
35. Škovránek, T., Podlubny, I., Petráš, I.: Modeling of the national economies in state-space: a fractional calculus approach. Econ. Model. **29**(4), 1322–1327 (2012)
36. Singh, J., Rashidi, M.M., Kumar, D., Swroop, R.: A fractional model of a dynamical Brusselator reaction-diffusion system arising in triple collision and enzymatic reactions. Nonlinear Eng. **5**(4), 277–285 (2016)
37. Al-Basir, F., Elaiw, A.M., Kesh, D., Roy, P.K.: Optimal control of a fractional-order enzyme kinetic model. Control Cybern. **44**(4), 443–461 (2015)
38. Gomez, F., Saad, K.: Coupled reaction-diffusion waves in a chemical system via fractional derivatives in Liouville-Caputo sense. Revista Mexicana de Física **64**(5), 539–547 (2018)
39. Isfer, L.A.D., Lenzi, M.K., Lenzi, E.K.: Identification of biochemical reactors using fractional differential equations. Lat. Am. Appl. Res. **40**(2), 193–198 (2010)
40. Singh, J., Kumar, D., Baleanu, D.: On the analysis of chemical kinetics system pertaining to a fractional derivative with Mittag-Leffler type kernel. Chaos Interdiscip. Nnlinear Sci. **27**(10), 103113 (2017)
41. Flores Tlacuahuac, A., Biegler, L.T.: Optimization of fractional order dynamic chemical processing systems. Ind. Eng. Chem. Res. **53**(13), 5110–5127 (2014)
42. Yadav, V.K., Das, S., Bhadauria, B.S., Singh, A.K., Srivastava, M.: Stability analysis, chaos control of a fractional order chaotic chemical reactor system and its function projective synchronization with parametric uncertainties. Chin. J. Phys. **55**(3), 594–605 (2017)
43. Tsirimokou, G., Laoudias, C., Psychalinos, C.: 0.5-V fractional-order companding filters. Int. J. Circuit Theory Appl. **43**(9), 1105–1126 (2015)
44. Chen, Y., Vinagre, B.M.: A new IIR-type digital fractional order differentiator. Signal Process. **83**(11), 2359–2365 (2003)
45. Tseng, C.C.: Design of FIR and IIR fractional order Simpson digital integrators. Signal Process. **87**(5), 1045–1057 (2007)
46. Maundy, B., Elwakil, A.S., Freeborn, T.J.: On the practical realization of higher-order filters with fractional stepping. Signal Process. **91**(3), 484–491 (2011)
47. Freeborn, T., Maundy, B., Elwakil, A.S.: Approximated fractional order Chebyshev lowpass filters. In: Mathematical Problems in Engineering (2015)
48. Radwan, A.G., Soliman, A.M., Elwakil, A.S.: Design equations for fractional-order sinusoidal oscillators: four practical circuit examples. Int. J. Circuit Theory Appl. **36**(4), 473–492 (2008)
49. Ali, A.S., Radwan, A.G., Soliman, A.M.: Fractional order Butterworth filter: active and passive realizations. IEEE J. Emerg. Sel. Top. Circuits Syst. **3**(3), 346–354 (2013)
50. Jesus, I.S., Machado, J.T.: Development of fractional order capacitors based on electrolyte processes. Nonlinear Dyn. **56**(1–2), 45–55 (2009)

51. Tsirimokou, G., Psychalinos, C., Elwakil, A.S., Salama, K.N.: Experimental verification of on-chip CMOS fractional-order capacitor emulators. Electron. Lett. **52**(15), 1298–1300 (2016)
52. Freeborn, T.J., Maundy, B., Elwakil, A.S.: Fractional-order models of supercapacitors, batteries and fuel cells: a survey. Mater. Renew. Sustain. Energy **4**(3), 9 (2015)
53. Tripathy, M.C., Mondal, D., Biswas, K., Sen, S.: Experimental studies on realization of fractional inductors and fractional-order bandpass filters. Int. J. Circuit Theory Appl. **43**(9), 1183–1196 (2015)
54. Galvõo, R.K.H., Hadjiloucas, S., Kienitz, K.H., Paiva, H.M., Afonso, R.J.M.: Fractional order modeling of large three-dimensional RC networks. IEEE Trans. Circuits Syst. I: Regul. Pap. **60**(3), 624–637 (2012)
55. Radwan, A.G., Salama, K.N.: Fractional-order RC and RL circuits. Circuits Syst. Signal Process. **31**(6), 1901–1915 (2012)
56. Adhikary, A., Sen, S., Biswas, K.: Practical realization of tunable fractional order parallel resonator and fractional order filters. IEEE Trans. Circuits Syst. I: Regul. Pap. **63**(8), 1142–1151 (2016)
57. Beltempo, A., Zingales, M., Bursi, O.S., Deseri, L.: A fractional-order model for aging materials: an application to concrete. Int. J. Solids Struct. **138**, 13–23 (2018)
58. Martynyuk, V., Ortigueira, M., Fedula, M., Savenko, O.: Methodology of electrochemical capacitor quality control with fractional order model. AEU-Int. J. Electron. Commun. **91**, 118–124 (2018)
59. Ahmad, W.M., El-Khazali, R.: Fractional-order dynamical models of love. Chaos Solitons Fractals **33**(4), 1367–1375 (2007)
60. Song, L., Xu, S., Yang, J.: Dynamical models of happiness with fractional order. Commun. Nonlinear Sci. Numer. Simul. **15**(3), 616–628 (2010)
61. Ahmed, E., El-Sayed, A.M.A., El-Saka, H.A.: Equilibrium points, stability and numerical solutions of fractional-order predator-prey and rabies models. J. Math. Anal. Appl. **325**(1), 542–553 (2007)
62. Singh, A.P., Agarwal, H., Srivastava, P., Naidu, Praveen: A robust fractional model predictive control design. Progress Fract. Differ. Appl. **5**(3), 217–223 (2019)
63. Srivastava, T., Singh, A.P., Agarwal, H.: Modeling the under-actuated mechanical system with fractional order derivative. Progress Fract. Differ. Appl. **1**(1), 57–64 (2015)
64. Dzieliński, A., Sarwas, G., Sierociuk, D.: Comparison and validation of integer and fractional order ultracapacitor models. Adv. Differ. Equ. **2011**(1), 11 (2011)
65. Manabe, S.: The non-integer integral and its application to control systems. J. Inst. Electr. Eng. Jpn **80**(860), 589–597 (1960)
66. Vinagre, B.M., Podlubny, I., Dorcak, L., Feliu, V.: On fractional PID controllers: a frequency domain approach. IFAC Proc. Vol. **33**(4), 51–56 (2000)
67. Caponetto, R., Fortuna, L., Porto, D.: Parameter tuning of a non integer order PID controller. In: 15th International Symposium on Mathematical Theory of Networks and Systems, Notre Dame, Indiana (2002)
68. Debbarma, S., Dutta, A.: Utilizing electric vehicles for LFC in restructured power systems using fractional order controller. IEEE Trans. Smart Grid **8**(6), 2554–2564 (2017)
69. Folea, S., Muresan, C.I., De Keyser, R., Ionescu, C.M.: Theoretical analysis and experimental validation of a simplified fractional order controller for a magnetic levitation system. IEEE Trans. Control Syst. Technol. **24**(2), 756–763 (2016)
70. Dimeas, I., Petras, I., Psychalinos, C.: New analog implementation technique for fractional-order controller: a DC motor control. AEU-Int. J. Electron. Commun. **78**, 192–200 (2017)
71. Nangrani, S.P., Bhat, S.S.: Fractional order controller for controlling power system dynamic behavior. Asian J. Control **20**(1), 403–414 (2018)
72. Deng, Z., Cao, H., Li, X., Jiang, J., Yang, J., Qin, Y.: Generalized predictive control for fractional order dynamic model of solid oxide fuel cell output power. J. Power Sources **195**(24), 8097–8103 (2010)
73. Domek, S.: Switched state model predictive control of fractional-order nonlinear discrete-time systems. Asian J. Control **15**(3), 658–668 (2013)

74. Sopasakis, P., Sarimveis, H.: Stabilising model predictive control for discrete-time fractional-order systems. Automatica **75**, 24–31 (2017)
75. Rhouma, A., Bouani, F., Bouzouita, B., Ksouri, M.: Model predictive control of fractional order systems. J. Comput. Nonlinear Dyn. **9**(3), 031011 (2014)
76. Oustaloup, A., Melchior, P., El Yagoubi, A.: A new control strategy based on the concept of non integer derivation: application in robot control. In: Information Control Problems in Manufacturing Technology 1989. Pergamon, pp. 641–648 (1990)
77. Machado, J.A.T.: Intelligent motion control using fractional integrals and derivatives. ASI'95, 1-9 (1995)
78. Machado, J.T., Azenha, A.: Fractional-order hybrid control of robot manipulators. In: SMC'98 Conference Proceedings. 1998 IEEE International Conference on Systems, Man, and Cybernetics (Cat. No. 98CH36218), vol. 1. IEEE, , pp. 788–793 (1998)
79. Duarte, F.B., Machado, J.T.: Fractional-order dynamics in the trajectory control of redundant manipulators. INES **2001**, 1–9 (2001)
80. Melchior, P., Orsoni, B., Lavialle, O., Oustaloup, A.: The CRONE toolbox for Matlab: fractional path planning design in robotics. In: Proceedings 10th IEEE International Workshop on Robot and Human Interactive Communication. ROMAN 2001 (Cat. No. 01TH8591). IEEE, pp. 534–540 (2001)
81. Duarte, F.B., Machado, J.T.: Chaotic phenomena and fractional-order dynamics in the trajectory control of redundant manipulators. Nonlinear Dyn. **29**(1–4), 315–342 (2002)
82. Duarte, F.B., Machado, J.T.: Pseudoinverse trajectory control of redundant manipulators: a fractional calculus perspective. In: Proceedings 2002 IEEE International Conference on Robotics and Automation (Cat. No. 02CH37292), vol. 3. IEEE, pp. 2406–2411 (2002)
83. Orsoni, B., Melchior, P., Oustaloup, A., Badie, T., Robin, G.: Fractional motion control: application to an XY cutting table. Nonlinear Dyn. **29**(1–4), 297–314 (2002)
84. Ferreira, N.F., Machado, J.T.: Fractional-order hybrid control of robotic manipulators. In: Proceedings 11th International Conference on Advanced Robotics, ICAR, vol. 1, pp. 393–398 (2003)
85. Pires, E.S., Machado, J.T., de Moura Oliveira, P.B.: Fractional order dynamics in a GA planner. Signal Process. **83**(11), 2377–2386 (2003)
86. Silva, M.F., Machado, J.T., Lopes, A.M.: Fractional order control of a hexapod robot. Nonlinear Dyn. **38**(1–4), 417–433 (2004)
87. Valerio, D., Sadacosta, J.: Fractional order control of a flexible robot. In: Fractional Differentiation and Its Applications, pp. 649–660 (2004)
88. Lazarevic, M.P.: Finite time stability analysis of PD^{α} fractional control of robotic time-delay systems. Mech. Res. Commun. **33**(2), 269–279 (2006)
89. Feliu, V., Vinagre, B.M., Monje, C.A.: Fractional-order control of a flexible manipulator. In: Advances in Fractional Calculus. Springer, Dordrecht, pp. 449–462 (2007)
90. Efe, M.: Fractional fuzzy adaptive sliding-mode control of a 2-DOF direct-drive robot arm. IEEE Trans. Syst. Man Cybern. Part B (Cybern.) **38**(6), 1561–1570 (2008)
91. Silva, M.F., Tenreiro Macahdo, J.A., Barbosa, R.S.: Using fractional derivatives in joint control of hexapod robots. J. Vib. Control **14**(9–10), 1473–1485 (2008)
92. Lima, M.F., Machado, J.A., Crisóstomo, M.: Fractional dynamics in mechanical manipulation. J. Comput. Nonlinear Dyn. **3**(2), 021203 (2008)
93. Monje, C.A., Vinagre, B.M., Feliu, V., Chen, Y.: Tuning and auto-tuning of fractional order controllers for industry applications. Control Eng. Pract. **16**(7), 798–812 (2008)
94. Marcos, M., Duarte, F.B., Machado, J.T.: Fractional dynamics in the trajectory control of redundant manipulators. Commun. Nonlinear Sci. Numer. Simul. **13**(9), 1836–1844 (2008)
95. Ferreira, N., Duarte, F., Lima, M., Marcos, M., Machado, J.: Application of fractional calculus in the dynamical analysis and control of mechanical manipulators. Fract. Calc. Appl. Anal. **11**(1), 91–113 (2008)
96. Couceiro, M.S., Ferreira, N.F., Machado, J.T.: Application of fractional algorithms in the control of a robotic bird. Commun. Nonlinear Sci. Numer. Simul. **15**(4), 895–910 (2010)

97. Delavari, H., Lanusse, P., Sabatier, J.: Fractional order controller design for a flexible link manipulator robot. Asian J. Control **15**(3), 783–795 (2013)
98. Dumlu, A., Erenturk, K.: Trajectory tracking control for a 3-dof parallel manipulator using fractional-order $PI^{\lambda}D^{\mu}$ control. IEEE Trans. Ind. Electron. **61**(7), 3417–3426 (2013)
99. Sharma, R., Rana, K.P.S., Kumar, V.: Performance analysis of fractional order fuzzy PID controllers applied to a robotic manipulator. Expert Syst. Appl. **41**(9), 4274–4289 (2014)
100. Mujumdar, A., Tamhane, B., Kurode, S.: Fractional order modeling and control of a flexible manipulator using sliding modes. In: 2014 American Control Conference. IEEE, pp. 2011–2016 (2014)
101. Sharma, R., Gaur, P., Mittal, A.P.: Performance analysis of two-degree of freedom fractional order PID controllers for robotic manipulator with payload. ISA Trans. **58**, 279–291 (2015)
102. Viola, J., Angel, L.: Identification, control and robustness analysis of a robotic system using fractional control. IEEE Lat. Am. Trans. **13**(5), 1294–1302 (2015)
103. Bensenouci, A., Shehata, M.: Optimized FOPID control of a single link flexible manipulator (SLFM) using genetic algorithm. In: Applied Mechanics and Materials, vol. 704. Trans Tech Publications, pp. 336–340 (2015)
104. Mohammed, R.H., Bendary, F., Elserafi, K.: Trajectory tracking control for robot manipulator using fractional order-fuzzy-PID controller. Int. J. Comput. Appl. **134**(15), 8887 (2016)
105. Sharma, R., Gaur, P., Mittal, A.P.: Design of two-layered fractional order fuzzy logic controllers applied to robotic manipulator with variable payload. Appl. Soft Comput. **47**, 565–576 (2016)
106. Fani, D., Shahraki, E.: Two-link robot manipulator using fractional order PID controllers optimized by evolutionary algorithms. Biosci. Biotechnol. Res. Asia **13**(1), 589–598 (2016)
107. Kumar, A., Kumar, V.: Hybridized ABC-GA optimized fractional order fuzzy pre-compensated FOPID Control design for 2-DOF robot manipulator. AEU-Int. J. Electron. Commun. **79**, 219–233 (2017)
108. Sharma, R., Gaur, P., Mittal, A.P.: Optimum design of fractional-order hybrid fuzzy logic controller for a robotic manipulator. Arab. J. Sci. Eng. **42**(2), 739–750 (2017)
109. Wang, L., Wang, D., Wu, J.: A control strategy of a 2-DOF parallel manipulator with fractional order PD^{μ} control. Int. J. Robot. Autom. **32**(4) (2017)
110. Asl, R.M., Pourabdollah, E., Salmani, M.: Optimal fractional order PID for a robotic manipulator using colliding bodies design. Soft Comput. **22**(14), 4647–4659 (2018)
111. Angel, L., Viola, J.: Fractional order PID for tracking control of a parallel robotic manipulator type delta. ISA Trans. **79**, 172–188 (2018)
112. Mohan, V., Chhabra, H., Rani, A., Singh, V.: An expert 2DOF fractional order fuzzy PID controller for nonlinear systems. Neural Comput. Appl. **31**(8), 4253–4270 (2019)
113. Singh, A.P., Deb, D., Agarwal, H.: On selection of improved fractional model and control of different systems with experimental validation. Commun. Nonlinear Sci. Numer. Simul. **79**, 104902 (2019)
114. Singh, A.P., Kazi, F.S., Singh, N.M., Srivastava, P.: $PI^{\alpha}D^{\beta}$ controller design for under actuated mechanical systems. In 2012 12th International Conference on control automation robotics & vision (ICARCV). IEEE, pp. 1654–1658 (2012)
115. Shah, P., Agashe, S.D., Singh, A.P.: Design of fractional order controller for undamped control system. In: 2013 Nirma University international conference on engineering (NUiCONE). IEEE, pp. 1–5 (2013)
116. Singh, A.P., Agarwal, H., Srivastava, P.: Fractional order controller design for inverted pendulum on a cart system (POAC). WSEAS Trans. Syst. Control **10**, 172–178 (2015)
117. Singh, A.P., Kazi, F., Singh, N.M., Vyawahare, V.: Fractional Order controller design for underactuated mechanical systems. In: The 5th IFAC Symposium on Fractional Differentiation and its Applications-FDA (2012)
118. Singh, A.P., Srivastava, T., Agrawal, H., Srivastava, P.: Fractional order controller design and analysis for crane system. Progress Fract. Differ. Appl. Int. J. **3**(2), 155–162 (2017)
119. Zhou, M., Feng, Y., Xue, C., Han, F.: Deep Convolutional neural network based fractional-order terminal sliding-mode control for robotic manipulators. In: Neurocomputing (2019)

120. Nojavanzadeh, D., Badamchizadeh, M.: Adaptive fractional-order non-singular fast terminal sliding mode control for robot manipulators. IET Control Theory Appl. **10**(13), 1565–1572 (2016)
121. Kumar, V., Rana, K.P.S.: Nonlinear adaptive fractional order fuzzy PID control of a 2-link planar rigid manipulator with payload. J. Frankl. Inst. **354**(2), 993–1022 (2017)
122. Deng, Y.: Fractional-order fuzzy adaptive controller design for uncertain robotic manipulators. Int. J. Adv. Robot. Syst. **16**(2), 1729881419840223 (2019)
123. Sharma, R., Bhasin, S., Gaur, P., Joshi, D.: A switching-based collaborative fractional order fuzzy logic controllers for robotic manipulators. Appl. Math. Model. **73**, 228–246 (2019)
124. Singh, A., Agrawal, H.: A fractional model predictive control design for 2-d gantry crane system. J. Eng. Sci. Technol. **13**(7), 2224–2235 (2018)
125. Singh, A.P., Agrawal, H., Srivastava, P.: Robust fractional model predictive controller (FMPC) design for under-actuated robotic systems. Int. J. Control Autom. **11**(7), 2224–2235 (2018)
126. Monje, C.A., Ramos, F., Feliu, V., Vinagre, B.M.: Tip position control of a lightweight flexible manipulator using a fractional order controller. IET Control Theory Appl. **1**(5), 1451–1460 (2007)
127. Nikdel, N., Badamchizadeh, M., Azimirad, V., Nazari, M.A.: Fractional-order adaptive backstepping control of robotic manipulators in the presence of model uncertainties and external disturbances. IEEE Trans. Ind. Electron. **63**(10), 6249–6256 (2016)
128. Wang, Y., Gu, L., Xu, Y., Cao, X.: Practical tracking control of robot manipulators with continuous fractional-order nonsingular terminal sliding mode. IEEE Trans. Ind. Electron. **63**(10), 6194–6204 (2016)
129. Mujumdar, A., Tamhane, B., Kurode, S.: Observer-based sliding mode control for a class of noncommensurate fractional-order systems. IEEE/ASME Trans. Mech. **20**(5), 2504–2512 (2015)
130. Wang, Y., Luo, G., Gu, L., Li, X.: Fractional-order nonsingular terminal sliding mode control of hydraulic manipulators using time delay estimation. J. Vib. Control **22**(19), 3998–4011 (2016)
131. Delavari, H., Azizkhani, A., Shiuooei, P.: Design and practical implementation of a fractional order pid controller for a single flexible-link robot. Modares Mech. Eng. **17**(10), 411–419 (2017)
132. Dwivedi, P., Pandey, S., Junghare, A.: Performance analysis and experimental validation of 2-DOF fractional-order controller for underactuated rotary inverted pendulum. Arab. J. Sci. Eng. **42**(12), 5121–5145 (2017)

Chapter 2
Fractional Modeling of Robotic Systems

A mathematical model helps represent the complex behavior of a dynamical system into a set of mathematical formulations to capture the essential characteristics and to characterize the impact of different constraints on the optimal system response. It helps to analyze the system response mathematically, perform simulations and provides a tool for testing theoretical phenomena. This chapter introduces mathematical modeling of robotic systems. The systems utilized for modeling are 2D Gantry Crane System, Double Inverted Pendulum on a Cart, Missile Launching Vehicle (MLV) or Pad System and Pendulum on a Cart. It is possible to extend this notion to many other systems such as ball and beam system, snake-type systems, locomotive systems, swimming robots, aircraft, acrobatic robots, helicopters, underwater vehicles, and surface vessels. Various methods of modeling are available such as Newtonian mechanics, Hamiltonian mechanics, Lagrangian mechanics. In this chapter, Euler-Lagrangian (E-L) formulation is used to model the system under consideration. These modeling equations are used for further analysis. In classical mechanics, problems related to the basic movements of objects are examined which has led to the designing of various fascinating models. There are numerous approaches which can be used to derive models from these essential systems. One of the most accepted approach is to utilize the Euler-Lagrange formulation [1, 2], for which the system energy needs to be calculated. The energy of a system can be classified as

(i) Potential Energy (PE), that is, stored energy; for example, when spring is compacted, or an object lifted at a certain height, and
(ii) Kinetic Energy (KE), which gets from the motion of the object.

Let P and V are potential and kinetic energy of the system; then the EL equation can be expressed as

$$\frac{d}{dt}\left(\frac{\partial(T-V)}{\partial\dot{x}}\right)-\frac{\partial(T-V)}{\partial x}=F, \tag{2.1}$$

A. P. Singh et al., *Fractional Modeling and Controller Design of Robotic Manipulators*, Intelligent Systems Reference Library 194,
https://doi.org/10.1007/978-3-030-58247-0_2

where x is position and v is velocity. To find the equations of motions for a given system, one must follow these steps:

(i) Find the values of PE and KE to compute the Lagrangian $L = T - V$.
(ii) Compute $\frac{dL}{dq}$.
(iii) Compute $\frac{dL}{d\dot{q}}$ and $\frac{d}{dt}\frac{dL}{d\dot{q}}$. It is important that $\dot{q}$ be treated as a complete variable in its own right, and not as a derivative.
(iv) Equate $\frac{dL}{dq} + F(t) = \frac{d}{dt}\frac{dL}{d\dot{q}}$.
(v) Solve the differential equation obtained in the preceding step. At this point, $\dot{q}$ is treated "normally". Note that the above equation might be a system of equations and not simply one equation.

2.1 Mathematical Modeling

Models portray how a particular system works [3, 4]. In mathematical modeling, we make a translation of those behaviors into equations due to the following merits:

- Mathematics helps understand system properties and basic assumptions.
- Mathematics is a compact language, with well-characterized rules for controls.
- For performing numerical calculations, PCs can be utilized.

There is a trade-off in mathematical modeling. The first form of trade-off is to distinguish the most significant elements of the system which should be incorporated into the system model, the rest are neglected [5]. The second form of trade-off is the complexity of the mathematical calculations. In the following sub sections a mathematical modeling of various robotic manipulators have been discussed.

2.1.1 Mass-Spring-Damper (MSD) System

Consider a simple mass-spring system Fig. 2.1 where a spring k is attached to the surface at one end and to a mass M at the opposite end. The force applied displaces the mass by $x(t)$.

The given system can be modeled by using either Newtonian mechanics or Lagrangian mechanics. First, consider the Newtonian mechanics to derive the modeling equation of the system. The applied vectored force on the system is equal and opposite to both the vector force of the spring and attached mass. It can be obtained from the FBD. From the FBD, we can obtain the vectored force applied which is equal and opposite of the vectored force of spring and vectored force on the attached mass:

$$\vec{F} = M\frac{d^2x}{dt^2} + kx. \tag{2.2}$$

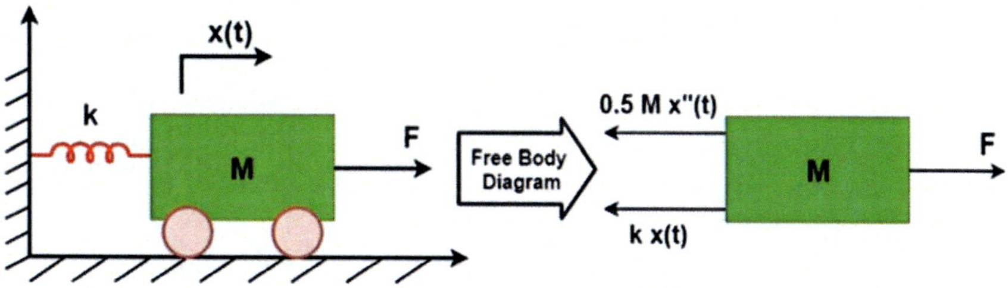

Fig. 2.1 Mass-spring system and its FBD

Now considering the Euler Lagrangian (EL) formulation to model the system, the Potential Energy (PE) and the Kinetic Energy (KE) of the system needs to be calculated. The PE of the system exists due to the spring only, and is given as

$$PE = \frac{1}{2}kx^2, \tag{2.3}$$

and KE of the system existing because of the attached mass, is given as

$$KE = \frac{1}{2}M(\dot{x})^2. \tag{2.4}$$

Then the Lagrangian will be

$$L = KE - PE = \frac{1}{2}M(\dot{x})^2 - \frac{1}{2}kx^2. \tag{2.5}$$

Putting this in to the EL equation, we get

$$M\ddot{x} + kx = F. \tag{2.6}$$

Both (2.2) and (2.6) represent the modeling equations of the same system using Newtonian mechanics and Lagrangian mechanics respectively. This book utilizes Lagrangian mechanics for modeling of the system. In Newtonian mechanics, it is needed to deal with vectors while in Lagrangian mechanics, only energies of the system are to be derived. Playing with energies is somewhat more comfortable than playing with vectors.

2.1.2 Inverted Pendulum on a Cart System (POAC)

Since the model of an inverted pendulum is nonlinear, we initially build up the necessary balanced conditions for the system, put these nonlinear conditions into a standard state, and then create a linearized model of the nonlinear plant. The

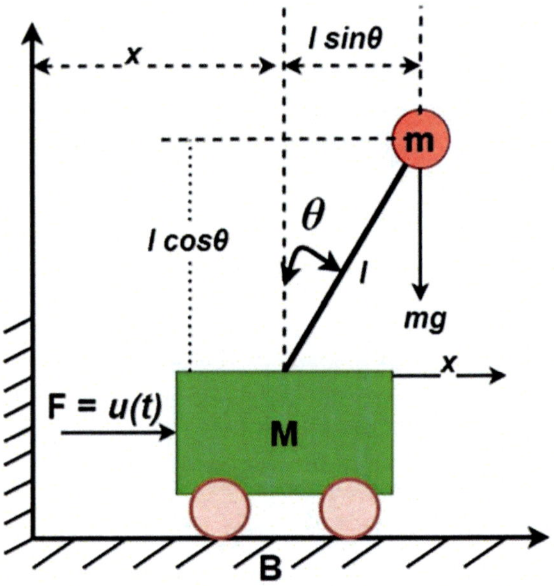

Fig. 2.2 Inverted pendulum on a cart

Lagrangian modeling analysis is used for the analysis of the considered system. The POAC system [6–9] appears in Fig. 2.2.

Initially, let us assume that the rod has zero mass and the cart mass is a point mass and are denoted by M and m respectively. A step force, i.e., a constant u(t) is applied on the cart in the direction shown in Fig. 2.2. We assume that B is the frictional coefficient and a gravity force follows up consistently, x(t) represents the position of the cart and the tilt angle is expressed by $\theta(t)$. Euler Lagrangian (EL) formulation is used for analysis and is given as

$$Ł = KE - PE, \tag{2.7}$$

for potential energy (PE) given as

$$PE = mgh = mgl\cos\theta, \tag{2.8}$$

where m represents point mass (kg), g represents acceleration (m/s^2), and $h = l\cos\theta$ is the height, and from the definition of kinetic energy, we have

$$KE = \frac{1}{2}mv^2, \tag{2.9}$$

where v is the velocity (m/s). The position vector of pendulum can be written as

$$\overrightarrow{r} = [l\sin\theta \quad l\cos\theta], \tag{2.10}$$

and for $v^2 = (\overrightarrow{\dot{r}})^2$ with $\overrightarrow{\dot{r}} = [l\cos\theta\dot{\theta} \quad -l\sin\theta\dot{\theta}]$, we get

$$(\overrightarrow{\dot{r}})^2 = (\overrightarrow{\dot{r}})\,(\overrightarrow{\dot{r}})^T = l^2\dot{\theta}^2, \tag{2.11}$$

such that the KE is

$$KE = \frac{1}{2}ml^2\dot{\theta}^2. \tag{2.12}$$

Therefore, from (2.7), we get

$$Ł = \frac{1}{2}ml^2\dot{\theta}^2 - mgl\cos\theta. \tag{2.13}$$

For a unit step input $F = u(t)$ provided to the pendulum, the Euler-Lagrange equation is given as

$$\frac{d}{dt}\left(\frac{\partial Ł}{\partial\dot{\theta}}\right) - \frac{\partial Ł}{\partial\theta} = u. \tag{2.14}$$

Using (2.13) and (2.14), we get

$$ml^2\ddot{\theta} - mgl\sin\theta = u, \tag{2.15}$$

which represents the nonlinear dynamical equation of an inverted pendulum.

Next, the modeling equations of POAC are derived, and for kinetic energy contributions from the cart (V_{cart}) and pendulum ($V_{pendulum}$) and potential energy contributions from the cart (T_{cart}) and pendulum ($T_{pendulum}$), the total potential and kinetic energies are given as

$$V_{total} = V_{cart} + V_{pendulum} = mgl\cos\theta, \tag{2.16}$$

$$T_{total} = T_{cart} + T_{pendulum} = \frac{1}{2}M\dot{x}^2 + \frac{1}{2}m(\dot{x}^2 + l^2\dot{\theta}^2 + 2\dot{x}\dot{\theta}l\cos\theta). \tag{2.17}$$

The Lagrangian is given as

$$Ł = T_{total} - V_{total} = \frac{1}{2}(M+m)\dot{x}^2 + \frac{1}{2}m(l^2\dot{\theta}^2 + 2\dot{x}\dot{\theta}l\cos\theta) - mgl\cos\theta, \tag{2.18}$$

and Rayleigh dissipation is expressed as

$$R = \frac{1}{2}B\ddot{x}^2. \tag{2.19}$$

Substituting (2.18) and (2.19) into EL formulation considering friction, we get

$$\frac{d}{dt}(\frac{\partial Ł}{\partial \dot{x}}) - \frac{\partial Ł}{\partial x} + \frac{\partial R}{\partial \dot{x}} = u, \tag{2.20}$$

and

$$\frac{d}{dt}(\frac{\partial Ł}{\partial \dot{\theta}}) - \frac{\partial Ł}{\partial \theta} = 0. \tag{2.21}$$

That is, we get

$$(M+m)\ddot{x} + ml\cos\theta\ddot{\theta} - ml\sin\theta\dot{\theta}^2 + B\dot{x} = u, \tag{2.22}$$

$$l\ddot{\theta} + \ddot{x}\cos\theta - g\sin\theta = 0, \tag{2.23}$$

and then the values of $\ddot{x}$ and $\ddot{\theta}$ can be found to be

$$\ddot{\theta} = \frac{(u\cos\theta + ml\sin\theta\cos\theta\dot{\theta}^2 - (M+m)s\sin\theta) - B\dot{x}\cos\theta}{ml\cos^2\theta - (M+m)l}, \tag{2.24}$$

$$\ddot{x} = \frac{(u + ml\sin\theta\dot{\theta}^2) - mg\sin\theta\cos\theta - B\dot{x}}{M+m-m\cos^2\theta}. \tag{2.25}$$

With $z_1 = x, z_2 = \dot{x} = \dot{z_1}, z_3 = \theta, z_4 = \dot{\theta} = \dot{z_3}$, we express in non-linear state-space form, and get

$$\frac{d}{dt}(\underline{z}) = \frac{d}{dt}\begin{pmatrix} z_1 \\ z_2 \\ z_3 \\ z_4 \end{pmatrix} = \underline{f}(\underline{z}, u, t) = \begin{pmatrix} z_2 \\ \frac{(u+ml\sin\theta\dot{\theta}^2)-mg\sin\theta\cos\theta}{M+m-m\cos^2\theta} \\ z_3 \\ \frac{(u\cos\theta+ml\sin\theta\cos\theta\dot{\theta}^2-(M+m)s\sin\theta)}{ml\cos^2\theta-(M+m)l} \end{pmatrix}. \tag{2.26}$$

Linearization provides the first order term representing the system [10]. Let us consider a system defined by

$$\frac{dx}{dt} = F(x,t), \tag{2.27}$$

then the linear system is written as

$$\frac{dx}{dt} = F(x_o,t) + DF(x_o,t).(x - x_o), \tag{2.28}$$

where x_o is the equilibrium point and $DF(x_o)$ is the Jacobian function of $F(x)$.

2.1.2.1 Linearization

Linearizing the state-space system obtained in (2.26) by evaluating the Jacobian at $(\underline{z_0}, u_0) = (0, 0)$ (for a detailed derivation, one may refer to [11]), we get

$$\frac{\partial f_i}{\partial z_j}, \tag{2.29}$$

for jth column at $(\underline{z_0}, u_0) = (0, 0)$, $j = 1, 2, 3, 4$, and therefore

$$\begin{pmatrix} \frac{df_1}{du} & \frac{df_2}{du} & \frac{df_3}{du} & \frac{df_4}{du} \end{pmatrix}^T, \tag{2.30}$$

for the input matrix at $(\underline{z_0}, u_0) = (0, 0)$. The system becomes

$$\frac{d}{dt}(\delta \underline{z}) = J_{=\underline{z}}(\underline{z_0}, u_0)\delta \underline{z} + J_{=u}(\underline{z_0}, u_0)\delta u. \tag{2.31}$$

Solving above, the following can be obtained,

$$\frac{d}{dt}(\delta \underline{z}) = \begin{pmatrix} 0 & 1 & 0 & 0 \\ 0 & \frac{-B}{M} & \frac{-mg}{M} & 0 \\ 0 & 0 & 0 & 1 \\ 0 & \frac{B}{Ml} & \frac{(M+m)g}{Ml} & 0 \end{pmatrix} \delta \underline{z} + \begin{pmatrix} 0 \\ \frac{1}{M} \\ 0 \\ \frac{-1}{Ml} \end{pmatrix} \delta u, \tag{2.32}$$

and the output matrix as,

$$y = \begin{pmatrix} 1 & 0 & 0 & 0 \\ 0 & 0 & 1 & 0 \end{pmatrix} \begin{pmatrix} x & \dot{x} & \theta & \dot{\theta} \end{pmatrix}^T. \tag{2.33}$$

2.1.2.2 Step Response of POAC System

In this section, a numerical simulation is presented to validate the linear model obtained for POAC system.

From Fig. 2.3, it can be observed that both the linear and nonlinear systems respond almost identically to a step input and hence the model obtained in (2.32) and (2.33) is a valid model because it is demonstrated that not much lost in going for the simpler linear system rather than the nonlinear system.

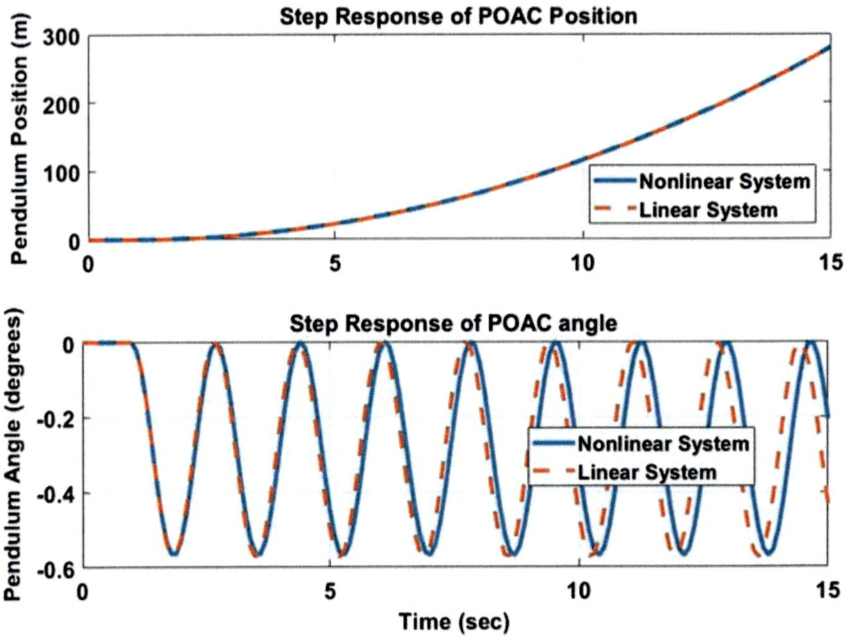

Fig. 2.3 Step response of POAC system

2.1.3 *Double Inverted Pendulum on a Cart System (DI-POAC)*

DI-POAC system [12–14] is shown in Fig. 2.4. Let us say that the rod has zero mass, cart mass is represented by M and the point masses are described by m_1 and m_2 individually.

A step force, i.e., a constant u(t) force is applied to the cart and the direction of the force is shown in Fig. 2.4. We assume that B is the frictional coefficient and force due to the gravity on the pendulum mass consistently, $x(t)$ represents the position of the cart, and the tilt angles are depicted in the figure by $\theta_1(t)$ and $\theta_2(t)$. For V_{total} and T_{total} as the total potential energy and total kinetic energy respectively, given as

$$V_{total} = V_{cart} + V_{pendulums} = m_1 g l_1 \cos\theta_1 + m_2 g\left(l_1 \cos\theta_1 + l_2 \cos\theta_2\right), \tag{2.34}$$

$$\begin{aligned} T_{total} = T_{cart} + T_{pendulums} &= \frac{1}{2}[M + m_1 + m_2]\dot{x}^2 + \frac{1}{2}[m_1 + m_2] l_1^2 \dot{\theta_1}^2 + \frac{1}{2} m_2 l_2^2 \dot{\theta_2}^2 \\ &+ [m_1 + m_2]\dot{x}\dot{\theta}_1 l_1 \cos\theta_1 + m_2 \dot{x}\dot{\theta}_2 l_2 \cos\theta_2 + l_1 l_2 \dot{\theta}_1 \dot{\theta}_2 \cos[\theta_1 - \theta_2], \end{aligned} \tag{2.35}$$

the Lagrangian can be expressed as

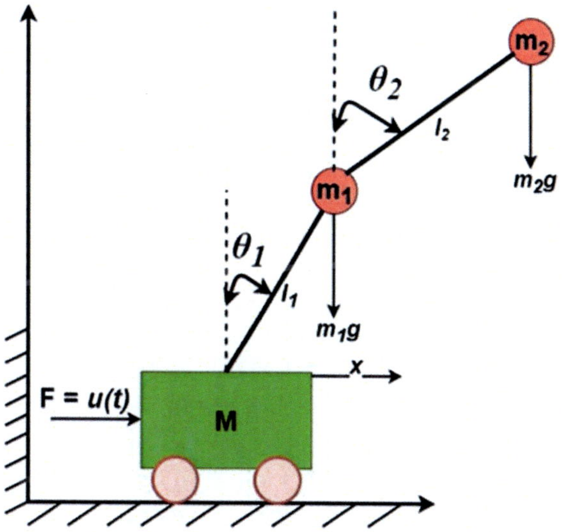

Fig. 2.4 Double inverted pendulum on a cart

$$
\begin{aligned}
Ł = T_{total} - V_{total} &= \frac{1}{2}[M + m_1 + m_2]\dot{x}^2 + \frac{1}{2}[m_1 + m_2]l_1^2\dot{\theta_1}^2 + \frac{1}{2}m_2 l_2^2\dot{\theta_2}^2 \\
&+[m_1 + m_2]\dot{x}\dot{\theta}_1 l_1 \cos\theta_1 + m_2\dot{x}\dot{\theta}_2 l_2 \cos\theta_2 + l_1 l_2 \dot{\theta}_1\dot{\theta}_2 \cos[\theta_1 - \theta_2] \\
&-[m_1 g l_1 \cos\theta_1 + m_2 g l_1 \cos\theta_1 + m_2 g l_2 \cos\theta_2]. \qquad (2.36)
\end{aligned}
$$

By substituting (2.36) into the EL equation

$$\frac{d}{dt}(\frac{\partial Ł}{\partial \dot{x}}) - \frac{\partial Ł}{\partial x} = u, \qquad (2.37)$$

we get

$$
\begin{aligned}
(M + m_1 + m_2)\ddot{x} + (m_1 + m_2)\ddot{\theta}_1 l_1 \cos\theta_1 + m_2\ddot{\theta}_2 l_2 \cos\theta_2 \\
- (m_1 + m_2)\dot{\theta_1}^2 l_1 \sin\theta_1 - m_2\dot{\theta_2}^2 l_2 \sin\theta_2 = u,
\end{aligned} \qquad (2.38)
$$

$$
\begin{aligned}
(m_1 + m_2)l_1^2\ddot{\theta_1} + (m_1 + m_2)\ddot{x} l_1 \cos\theta_1 + l_1 l_2 \ddot{\theta_2} \cos(\theta_1 - \theta_2) \\
+ l_1 l_2 \dot{\theta_2}^2 \sin(\theta_1 - \theta_2) - (m_1 + m_2) g l_1 \sin\theta_1 = 0,
\end{aligned} \qquad (2.39)
$$

$$
\begin{aligned}
m_2 l_2^2 \ddot{\theta_2} + m_2\ddot{x} l_2 \cos\theta_2 + l_1 l_2 \ddot{\theta_1} \cos(\theta_1 - \theta_2) \\
- l_1 l_2 \dot{\theta_1}^2 \sin(\theta_1 - \theta_2 + m_2 g l_2 \sin\theta_2) = 0.
\end{aligned} \qquad (2.40)
$$

Defining state variables $\underline{z} = [z_1 \ z_2 \ z_3 \ z_4 \ z_5 \ z_6]^T$ such that $z_1 = x$, $z_2 = \dot{x} = \dot{z_1}$, $z_3 = \theta_1$, $z_4 = \dot{\theta_1} = \dot{z_3}$, $z_5 = \theta_2$, $z_6 = \dot{\theta_2} = \dot{z_5}$, we formulate a state-space system as

$$\frac{d}{dt}(\underline{z}) = \underline{f}(\underline{z}, u, t). \tag{2.41}$$

By linearizing (2.41) and choosing the following values $m_1 = 0.25\,\text{Kg}$, $m_2 = 0.25\,\text{Kg}$, $M = 0.5\,\text{Kg}$, $l_1 = 1\,\text{m}$, $l_2 = 1\,\text{m}$, and $g = 10\,(\frac{\text{m}}{\text{s}^2})$, the linearized system representation in state-space can be obtained using

$$\frac{d}{dt}(\delta\underline{z}) = J_{=\underline{z}}(\underline{z_0}, u_0)\delta\underline{z} + J_{=u}(\underline{z_0}, u_0)\delta u \tag{2.42}$$

and the linearised state-space is given as

$$\frac{d}{dt}(\delta\underline{z}) = \begin{pmatrix} 0 & 1 & 0 & 0 & 0 & 0 \\ 0 & 0 & -0.073 & 0 & 0.12 & 0 \\ 0 & 0 & 0 & 1 & 0 & 0 \\ 0 & 0 & -1.42 & 0 & -2.873 & 0 \\ 0 & 0 & 0 & 0 & 0 & 1 \\ 0 & 0 & 5.75 & 0 & 1.382 & 0 \end{pmatrix} \delta\underline{z} + \begin{pmatrix} 0 \\ 0.102 \\ 0 \\ -0.015 \\ 0 \\ -0.015 \end{pmatrix} \delta u, \tag{2.43}$$

and the output matrix as

$$y = \begin{pmatrix} 1 & 0 & 0 & 0 & 0 & 0 \\ 0 & 0 & 1 & 0 & 0 & 0 \\ 0 & 0 & 0 & 0 & 1 & 0 \end{pmatrix} \begin{pmatrix} x & \dot{x} & \theta_1 & \dot{\theta}_1 & \theta_2 & \dot{\theta}_2 \end{pmatrix}^T. \tag{2.44}$$

2.1.4 2D-Gantry Crane System

The 2D-Gantry crane system [15–17] is generally used in large scale industries where it is required to pick and place heavyweight objects. The schematic is shown in Fig. 2.5. The cranes generally handle heavy weight and can be seen at depots, factories and ships, etc. Efficient control of such types of systems are needed and so this example is opted in this section. Modeling equations of 2-D Gantry crane system are derived by assuming the following:

(i) The load swing is limited to a few degrees to satisfy linear system properties.
(ii) The system starts at equilibrium with zero initial conditions.
(iii) Cable length is constant, and the weight of the cable is negligible.

Let m refers to the load mass and M denotes the trolley mass, l denotes the length of the cable, $u(t)$ is the force applied, $m\,g$ is the force of the gravity acting constantly, and $x(t)$ is the position of the trolley and $\theta(t)$ is the swing angle.

For controller design, the system model equations are needed. We used Euler-Lagrangian (EL) formulation to model the previous systems. From Fig. 2.5, one can

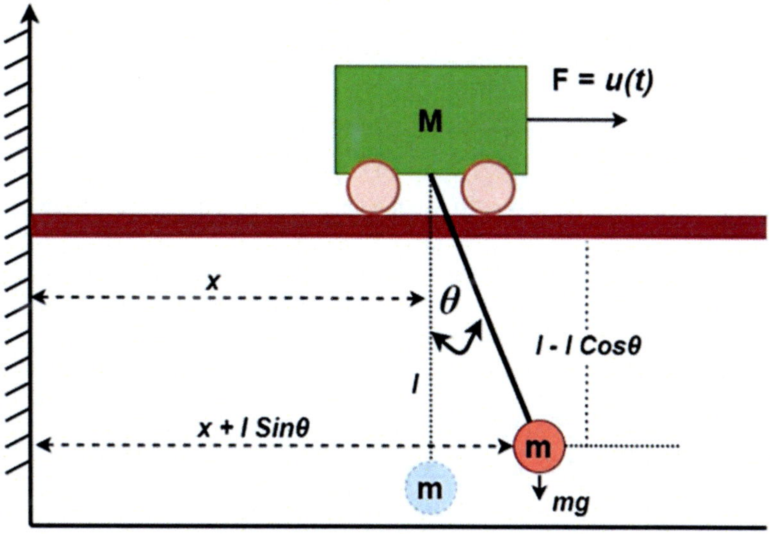

Fig. 2.5 Schematic diagram of 2-D gantry crane

find the position projections of the load as $(x + lSin\theta, -lCos\theta)$. So, for the same variable definitions as the earlier systems, we have

$$
\begin{aligned}
V_{total} &= V_{trolly} + V_{load} = Mgl + mg(l - l\cos\theta), \\
T_{total} &= T_{trolly} + T_{load} = \frac{1}{2}M\dot{x}^2 + \frac{1}{2}m(\dot{x}^2 + l^2\dot{\theta}^2 + 2\dot{x}\dot{\theta}l\cos\theta),
\end{aligned} \tag{2.45}
$$

and so the Lagrangian of the system is

$$
L = \frac{1}{2}M\dot{x}^2 + \frac{1}{2}m(\dot{x}^2 + l^2\dot{\theta}^2 + 2\dot{x}\dot{\theta}l\cos\theta) - (Mgl + mg(l - l\cos\theta)) \tag{2.46}
$$

Formulating the Lagrangian L from (2.46) to EL equations

$$
\frac{d}{dt}(\frac{\partial L}{\partial \dot{x}}) - \frac{\partial L}{\partial x} = u, \quad \frac{d}{dt}(\frac{\partial L}{\partial \dot{\theta}}) - \frac{\partial L}{\partial \theta} = 0, \tag{2.47}
$$

we get

$$
(M + m)\ddot{x} + ml\cos\theta\ddot{\theta} - ml\sin\theta\dot{\theta}^2 = u, \quad l\ddot{\theta} + \ddot{x}\cos\theta + g\sin\theta = 0, \tag{2.48}
$$

which when cast into nonlinear state-space form with state variables $z_1 = x$, $z_2 = \dot{x} = \dot{z_1}$, $z_3 = \theta$, $z_4 = \dot{\theta} = \dot{z_3}$, is given as

$$\frac{d}{dt}(\underline{z}) = \frac{d}{dt}\begin{pmatrix} z_1 \\ z_2 \\ z_3 \\ z_4 \end{pmatrix} = \begin{pmatrix} z_2 \\ \frac{(u+ml\sin z_3 z_4{}^2)+mg\sin z_3\cos z_3}{M+m-m\cos^2 z_3} \\ z_3 \\ \frac{(u\cos z_3+ml\sin z_3\cos z_3 z_4{}^2+(M+m)g\sin\theta)}{ml\cos^2 z_3-(M+m)l} \end{pmatrix}. \tag{2.49}$$

Talking Jacobian at $(\underline{z_0}, u_0) = (0, 0)$, the linearised matrix takes the form

$$\frac{d}{dt}(\delta\underline{z}) = J_{=\underline{z}}(\underline{z_0}, u_0)\delta\underline{z} + J_{=u}(\underline{z_0}, u_0)\delta u = \begin{pmatrix} 0 & 1 & 0 & 0 \\ 0 & 0 & \frac{mg}{M} & 0 \\ 0 & 0 & 0 & 1 \\ 0 & 0 & \frac{-(M+m)g}{Ml} & 0 \end{pmatrix}\delta\underline{z} + \begin{pmatrix} 0 \\ \frac{1}{M} \\ 0 \\ \frac{-1}{Ml} \end{pmatrix}\delta u, \tag{2.50}$$

and the output is

$$y = \begin{pmatrix} 1 & 0 & 0 & 0 \\ 0 & 0 & 1 & 0 \end{pmatrix}\begin{pmatrix} x & \dot{x} & \theta & \dot{\theta} \end{pmatrix}^T. \tag{2.51}$$

2.1.5 *Missile Launching Vehicle/Pad (MLV)*

The central aim for analysis of certain systems is in finding the governing equations. The system considered here is the MLV system [18] with two degrees of freedom: one angular motion in the vertical direction and the other angular motion in the horizontal direction. The schematic consisting of two serial links is shown in Fig. 2.6. The Link one has the axis of rotation as the y axis and the second link has the axis of rotation as the Z axis. Let l_1 and l_2 be the link lengths, $m_i, i = 1, 2$ be the mass placed on link i. Let θ_1 be the angle obtained through rotating link 1 in y axis and θ_2 is the angle obtained through rotating link 2 in z axis and g is the gravity force.

Fig. 2.6 Geometrical model of missile launch vehicle (MLV) System

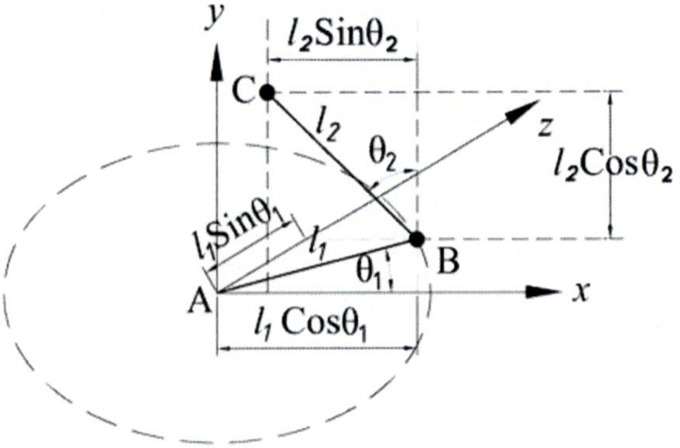

EL formulation for deriving governing motion equations of the MLV system [19] are

$$\frac{d}{dt}(\frac{\partial L}{\partial \dot{x}}) - \frac{\partial L}{\partial x} = F, \tag{2.52}$$

where x is the position, $\dot{x}$ is the velocity, F is the external force which is acting on the system. The position, and the rate of change of position of mass m_1 are given as

$$r_1 = \quad \begin{pmatrix} l_1 \cos\theta_1 & 0 & l_1 \sin\theta_1 \end{pmatrix}, \vec{\dot{r_1}} = \begin{pmatrix} -l_1\dot{\theta}_1 \sin\theta_1 & 0 & l_1\dot{\theta}_1 \cos\theta_1 \end{pmatrix}, \tag{2.53}$$

and therefore

$$\vec{\dot{r_1}}^2 = \begin{pmatrix} -l_1\dot{\theta}_1 \sin\theta_1 \; 0 \; l_1\dot{\theta}_1 \cos\theta_1 \end{pmatrix} \begin{pmatrix} -l_1\dot{\theta}_1 \sin\theta_1 \\ 0 \\ l_1\dot{\theta}_1 \cos\theta_1 \end{pmatrix} = {l_1}^2\dot{\theta}_1^2. \tag{2.54}$$

Similarly,

$$\vec{r_2} = \begin{pmatrix} l_1 \cos\theta_1 - l_2 \sin\theta_2 & l_2 \cos\theta_2 & l_1 \sin\theta_1 - l_2 \sin\theta_2 \end{pmatrix}, \tag{2.55}$$

$$\vec{\dot{r_2}} = \begin{pmatrix} -l_1\dot{\theta}_1 \sin\theta_1 - l_2\dot{\theta}_2 \cos\theta_2 & -l_2\dot{\theta}_2 \sin\theta_2 & l_1\dot{\theta}_1 \cos\theta_1 - l_2\dot{\theta}_2 \cos\theta_2 \end{pmatrix},$$

$$\begin{aligned} \vec{\dot{r_2}}^2 &= \begin{pmatrix} -l_1\dot{\theta}_1 \sin\theta_1 - l_2\dot{\theta}_2 \cos\theta_2 \; -l_2\dot{\theta}_2 \sin\theta_2 \; l_1\dot{\theta}_1 \cos\theta_1 - l_2\dot{\theta}_2 \cos\theta_2 \end{pmatrix} \\ &\quad \cdot \begin{pmatrix} -l_1\dot{\theta}_1 \sin\theta_1 - l_2\dot{\theta}_2 \cos\theta_2 \\ -l_2\dot{\theta}_2 \sin\theta_2 \\ l_1\dot{\theta}_1 \cos\theta_1 - l_2\dot{\theta}_2 \cos\theta_2 \end{pmatrix}, \\ &= {l_1}^2\dot{\theta}_1^2 + {l_2}^2\dot{\theta}_2^2 + {l_2}^2\dot{\theta}_2^2 \cos^2\theta_2 + 2l_1l_2\dot{\theta}_1\dot{\theta}_2(\sin\theta_1 - \cos\theta_1)\cos\theta_2. \end{aligned} \tag{2.56}$$

The total Potential Energy (PE) is expressed as

$$PE = PE_1(of\ mass\ m_1) + PE_2(of\ mass\ m_2) = m_2 g l_2 \cos\theta_2. \tag{2.57}$$

The total Kinetic Energy KE is

$$\begin{aligned} KE &= KE_1(of\ mass\ m_1) + KE_2(of\ mass\ m_2) \\ &= \frac{1}{2}m_1{l_1}^2\dot{\theta}_1^2 + \frac{1}{2}m_2\left({l_1}^2\dot{\theta}_1^2 + {l_2}^2\dot{\theta}_2^2(1 + \cos^2\theta_2)\right. \\ &\quad \left. + 2l_1l_2\dot{\theta}_1\dot{\theta}_2(\sin\theta_1 - \cos\theta_1)\cos\theta_2\right). \end{aligned} \tag{2.58}$$

So, Lagrangian L is

$$
\begin{aligned}
L = &\frac{1}{2}m_1 {l_1}^2 \dot{\theta}_1^2 - m_2 g l_2 \cos\theta_2 \\
&+\frac{1}{2}m_2\left({l_1}^2\dot{\theta}_1^2 + {l_2}^2\dot{\theta}_2^2(1+\cos^2\theta_2) + 2l_1l_2\dot{\theta}_1\dot{\theta}_2(\sin\theta_1 - \cos\theta_1)\cos\theta_2\right).
\end{aligned}
\tag{2.59}
$$

Now expressing (2.59) in the EL formulation

$$
\frac{d}{dt}\left(\frac{\partial L}{\partial \dot{\theta}_i}\right) - \frac{\partial L}{\partial \theta_i} = \tau_i,\ i = 1, 2, \tag{2.60}
$$

we get the inputs provided at the joints, given as

$$
\begin{aligned}
\tau_1 = &(m_1 + m_2){l_1}^2\ddot{\theta}_1 + m_2l_1l_2(\ddot{\theta}_2(\sin\theta_1\cos\theta_2 - \cos\theta_1\cos\theta_2)) \\
&+\dot{\theta}_2^2(\cos\theta_1\sin\theta_2 - \sin\theta_1\sin\theta_2), \\
\tau_2 = &\frac{1}{2}{l_2}^2m_2\ddot{\theta}_2 + \frac{1}{2}{l_2}^2m_2\ddot{\theta}_2\cos 2\theta_2 - \frac{1}{2}{l_2}^2m_2\dot{\theta}_2^2\sin 2\theta_2 \\
&+m_2l_1l_2\left(\ddot{\theta}_1(\sin\theta_1\cos\theta_2 - \cos\theta_1\cos\theta_2) + \dot{\theta}_1^2(\cos\theta_1\cos\theta_2 - \sin\theta_1\cos\theta_2)\right) \\
&+{l_2}^2m_2\ddot{\theta}_2 - m_2gl_2\sin\theta_2.
\end{aligned}
\tag{2.61}
$$

Expressing in terms of taking out $\ddot{\theta}_1$ and $\ddot{\theta}_2$, we get

$$
\begin{aligned}
\ddot{\theta}_1 &= \frac{E\cdot A + m_2l_1l_2(\sin\theta_1\cos\theta_2 - \cos\theta_1\cos\theta_2)\cdot B}{C}, \\
\ddot{\theta}_2 &= \frac{m_2l_1l_2(\sin\theta_1\cos\theta_2 - \cos\theta_1\cos\theta_2)\cdot A - (m_1+m_2){l_1}^2\cdot B}{D},
\end{aligned}
\tag{2.62}
$$

where

$$
\begin{aligned}
A &= \tau_1 - m_2l_1l_2\dot{\theta}_2^2(\sin\theta_2\cos\theta_1 - \sin\theta_1\sin\theta_2), \\
B &= \tau_2 + m_2gl_2\sin\theta_2 + \frac{1}{2}{l_2}^2m_2\dot{\theta}_2^2\sin 2\theta_2 - m_2l_1l_2\ddot{\theta}_1^2(\cos\theta_1\cos\theta_2 - \sin\theta_1\cos\theta_2), \\
E &= \frac{1}{2}{l_2}^2m_2(3+\cos 2\theta_2), \\
C &= E(m_1+m_2){l_1}^2 - {m_2}^2{l_1}^2{l_2}^2(\sin\theta_1\cos\theta_2 - \cos\theta_1\cos\theta_2)^2, \\
D &= {m_2}^2{l_1}^2{l_2}^2(\sin\theta_1 - \cos\theta_1)^2\cos^2\theta_2 + \frac{1}{2}(m_1+m_2)l_1^2l_2^2m_2\,(3+\cos 2\theta_2).
\end{aligned}
\tag{2.63}
$$

Now, presenting (2.62) into nonlinear standard state space equation, we have

$$
\frac{d}{dt}(\underline{z}) = \underline{f}(\underline{z}, u, t). \tag{2.64}
$$

Let $z_1 = \theta_1$, $z_2 = \dot{\theta}_1 = \dot{z_1}$, $z_3 = \theta_2$, $z_4 = \dot{\theta}_2 = \dot{z_3}$ and consider $l_2 = l_1 = 1\,\text{m}$, $m_2 = 20\,\text{Kg}$, $m_1 = 0.5\,\text{Kg}$ and $g = 9.8\,\text{m/s}^2$ for linearisation.

A linear model can be found to be

$$\frac{d}{dt}(\delta \underline{z}) = A \begin{pmatrix} z_1 \\ z_2 \\ z_3 \\ z_4 \end{pmatrix} + B \begin{pmatrix} \tau_1 \\ \tau_2 \end{pmatrix}, \tag{2.65}$$

where A and B matrices are summarized in Table 2.1 depending on initial conditions.

The output equation can be found to be

$$y = \begin{pmatrix} 1\ 0\ 0\ 0 \\ 0\ 0\ 1\ 0 \end{pmatrix} \begin{pmatrix} z_1 & z_2 & z_3 & z_4 \end{pmatrix}^T. \tag{2.66}$$

Table 2.1 Linear model of MLV depending on initial conditions

Model	Initial conditions	Matrix [A]	Matrix [B]
1	(0,0,0,0)	$\begin{pmatrix} 0 & 1 & 0 & 0 \\ 0 & 0 & -9.61 & 0 \\ 0 & 0 & 0 & 1 \\ 0 & 0 & -1.63 & 0 \end{pmatrix}$	$\begin{pmatrix} 0 & 0 \\ 0.1 & -0.02 \\ 0 & 0 \\ -0.01 & -0.004 \end{pmatrix}$
2	$(0, 0, pi/4, 0)$	$\begin{pmatrix} 0 & 1 & 0 & 0 \\ 9.6 & 0 & -0.05 & 0 \\ 0 & 0 & 0 & 1 \\ -0.86 & 0 & -3.5 & 0 \end{pmatrix}$	$\begin{pmatrix} 0 & 0 \\ 0.07 & -0.02 \\ 0 & 0 \\ -0.01 & -0.01 \end{pmatrix}$
3	$(pi/6, 0, pi/4, 0)$	$\begin{pmatrix} 0 & 1 & 0 & 0 \\ 5.05 & 0 & -0.75 & 0 \\ 0 & 0 & 0 & 1 \\ -0.7 & 0 & -3.8 & 0 \end{pmatrix}$	$\begin{pmatrix} 0 & 0 \\ 0.05 & -0.005 \\ 0 & 0 \\ -0.004 & -0.01 \end{pmatrix}$
4	$(pi/6, 0, pi/4, 0)$	$\begin{pmatrix} 0 & 1 & 0 & 0 \\ 5.05 & 0 & -0.75 & 0 \\ 0 & 0 & 0 & 1 \\ -0.7 & 0 & -3.8 & 0 \end{pmatrix}$	$\begin{pmatrix} 0 & 0 \\ 0.05 & -0.005 \\ 0 & 0 \\ -0.004 & -0.01 \end{pmatrix}$
5	$(pi/3, 0, pi/4, 0)$	$\begin{pmatrix} 0 & 1 & 0 & 0 \\ 5.05 & 0 & 0.75 & 0 \\ 0 & 0 & 0 & 1 \\ 0.7 & 0 & -3.8 & 0 \end{pmatrix}$	$\begin{pmatrix} 0 & 0 \\ 0.05 & 0.005 \\ 0 & 0 \\ 0.004 & -0.01 \end{pmatrix}$
6	$(pi/2, 0, pi/4, 0)$	$\begin{pmatrix} 0 & 1 & 0 & 0 \\ 9.6 & 0 & 0.05 & 0 \\ 0 & 0 & 0 & 1 \\ 0.9 & 0 & -3.5 & 0 \end{pmatrix}$	$\begin{pmatrix} 0 & 0 \\ 0.07 & 0.02 \\ 0 & 0 \\ 0.01 & -0.01 \end{pmatrix}$

(continued)

Table 2.1 (continued)

Model	Initial conditions	Matrix [A]	Matrix [B]
7	$(2pi/3, 0, pi/4, 0)$	$\begin{pmatrix} 0 & 1 & 0 & 0 \\ 13 & 0 & -17 & 0 \\ 0 & 0 & 0 & 1 \\ 0.3 & 0 & -3.1 & 0 \end{pmatrix}$	$\begin{pmatrix} 0 & 0 \\ 0.13 & 0.04 \\ 0 & 0 \\ 0.01 & -0.01 \end{pmatrix}$
8	$(5pi/6, 0, pi/4, 0)$	$\begin{pmatrix} 0 & 1 & 0 & 0 \\ -13 & 0 & -17 & 0 \\ 0 & 0 & 0 & 1 \\ -0.3 & 0 & -3.1 & 0 \end{pmatrix}$	$\begin{pmatrix} 0 & 0 \\ 0.13 & 0.04 \\ 0 & 0 \\ 0.01 & -0.01 \end{pmatrix}$
9	$(pi, 0, pi/4, 0)$	$\begin{pmatrix} 0 & 1 & 0 & 0 \\ -9.6 & 0 & 0.05 & 0 \\ 0 & 0 & 0 & 1 \\ -0.9 & 0 & -3.5 & 0 \end{pmatrix}$	$\begin{pmatrix} 0 & 0 \\ 0.1 & 0.02 \\ 0 & 0 \\ 0.01 & -0.01 \end{pmatrix}$
10	$(7pi/6, 0, pi/4, 0)$	$\begin{pmatrix} 0 & 1 & 0 & 0 \\ -5.05 & 0 & 0.75 & 0 \\ 0 & 0 & 0 & 1 \\ -0.7 & 0 & -3.8 & 0 \end{pmatrix}$	$\begin{pmatrix} 0 & 0 \\ 0.05 & 0.005 \\ 0 & 0 \\ 0.004 & -0.01 \end{pmatrix}$
11	$(4pi/3, 0, pi/4, 0)$	$\begin{pmatrix} 0 & 1 & 0 & 0 \\ -5.05 & 0 & 0.75 & 0 \\ 0 & 0 & 0 & 1 \\ 0.7 & 0 & -3.8 & 0 \end{pmatrix}$	$\begin{pmatrix} 0 & 0 \\ 0.05 & -0.005 \\ 0 & 0 \\ -0.004 & -0.01 \end{pmatrix}$
12	$(3pi/2, 0, pi/4, 0)$	$\begin{pmatrix} 0 & 1 & 0 & 0 \\ -9.6 & 0 & -0.05 & 0 \\ 0 & 0 & 0 & 1 \\ 0.9 & 0 & -3.5 & 0 \end{pmatrix}$	$\begin{pmatrix} 0 & 0 \\ 0.07 & -0.02 \\ 0 & 0 \\ -0.01 & -0.01 \end{pmatrix}$
13	$(5pi/3, 0, pi/4, 0)$	$\begin{pmatrix} 0 & 1 & 0 & 0 \\ -13 & 0 & 17 & 0 \\ 0 & 0 & 0 & 1 \\ 0.3 & 0 & -3.1 & 0 \end{pmatrix}$	$\begin{pmatrix} 0 & 0 \\ 0.13 & -0.04 \\ 0 & 0 \\ -0.01 & -0.01 \end{pmatrix}$
14	$(11pi/6, 0, pi/4, 0)$	$\begin{pmatrix} 0 & 1 & 0 & 0 \\ 13 & 0 & 17 & 0 \\ 0 & 0 & 0 & 1 \\ -0.3 & 0 & -3.1 & 0 \end{pmatrix}$	$\begin{pmatrix} 0 & 0 \\ 0.13 & -0.04 \\ 0 & 0 \\ -0.01 & -0.01 \end{pmatrix}$

2.2 Fractional Modeling of a System

Before setting out to model a system, the objectives should be clear. Then we develop the essential equations of the model which reflect the functioning of the system. These pieces of information can be expressed through some assumptions. In the event that the assumptions are adequate, they may lead straightforwardly to the mathematical equations governing the system.

We live in an analog fractional world and fractional order systems analysis is central to developing more ideal control schemes and understanding of the world. The fractional calculus concept is ideal but it is more interesting, the other interesting thing is that even though the fractional effects we are seeing research interest but they are partially successful. Fractional calculus is calculus with fractional derivatives. The idea is that from ordinary calculus we know that we have a first order derivative which is velocity and then second derivative which is acceleration, it turns out that we can have a derivative of any order between 0 and 1 or 1 and 2 like wise, in fact any order at all. And the interesting this is that these were discovered by Leibniz in1695 a few years after he discovered ordinary calculus. But they were forgotten because the formulas to evaluate those fractional derivatives are so complex as it's almost impossible to work with them using pencil and paper. So no wonder they were forgotten then, but now we have computers and different algorithms and the complexity is no longer a problem.

In this chapter, fractional embedding to the robotic systems is analyzed using the definition of Laplace transform of FO systems, as represented in [20, 21]. The Laplace transform of such a system is shown below

$$L[Dx(t) - 3D^{\frac{1}{2}}x(t) + 2D^{0}x(t) = 0], \tag{2.67}$$

$$sX(s) - x(0) - 3s^{\frac{1}{2}}X(s) + 3D^{-\frac{1}{2}}x(0) + 2x(s) = 0. \tag{2.68}$$

Different systems have been considered in this chapter. The model equations derived earlier are the necessary foundation for this section. The FO model derivation uses the concept of FO Laplace transformation with zero initial conditions. Fractional order system modeling results in a better performance compared to the Integer model. Implementation of Fractional model is comparatively complex, but it results in improved performance like robustness, transient stability, noise filtering and disturbance rejection, etc. [22]. Designing an optimized controller for a given system is another major issue, and FO controllers have been demonstrated to be efficient. Fractional order PID (FOPID) controllers were first mentioned in Podlubny's work in which, for a FO system, a FO integrator and a FO differentiator is designed.

We choose a simple RL circuit shown in Fig. 2.7 which consists of a resistor (R) in series with an inductor (L), to understand fractional modeling. This series combination is then connected to a supply voltage through a switch. At $t = 0$, the switch is closed, which makes the initial conditions of the circuit parameters equal to zero ($i(0) = 0$) because the inductor doesn't allow sudden change in current.

Using KVL (Kirchhoff's Voltage Law), we have

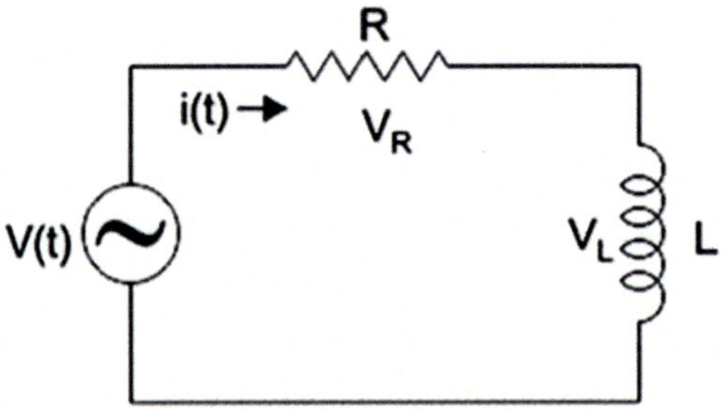

Fig. 2.7 RL circuit

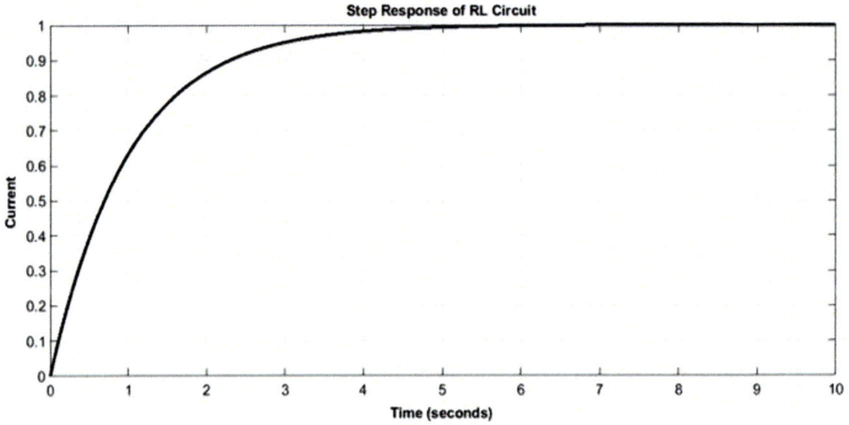

Fig. 2.8 Step response of RL circuit

$$V(t) - i(t)R - L\frac{di(t)}{dt} = 0. \tag{2.69}$$

Taking the Laplace transform and assuming $R = 1\Omega$, $H = 1$ Henry, we get

$$\frac{I(s)}{V(s)} = \frac{1}{s+1}, \tag{2.70}$$

and the step response of (2.70) (in Amperes) is

$$i(t) = 1 - e^{-t}, \tag{2.71}$$

for $t > 0$. Figure 2.8 depicts the response of this system.

The equations from (2.69) to (2.71) are in the integer domain. Let us do the analysis of these equations in fractional domain. Taking Eq. (2.69) and writing its fractional equivalent, we get

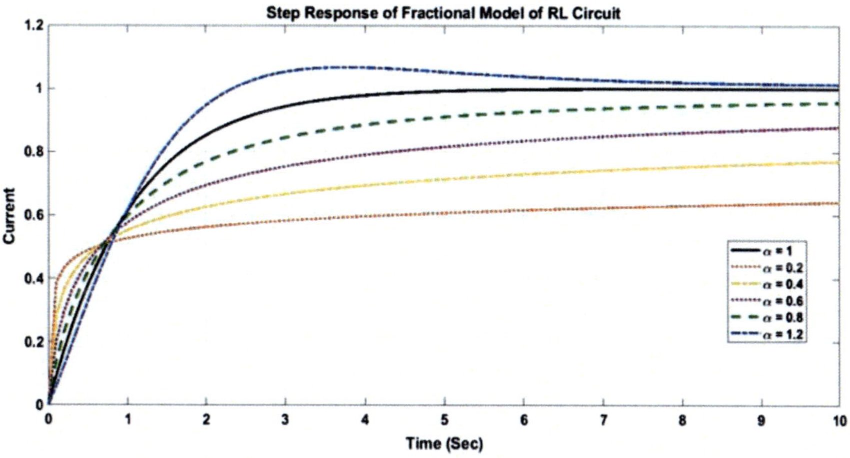

Fig. 2.9 Step response of fractional order model of RL circuit

$$V(t) - i(t)R - L\frac{d^{\alpha}i(t)}{dt^{\alpha}} = 0. \tag{2.72}$$

Clearly for $\alpha = 1$ in (2.72) we get back (2.69). The Laplace transform of (2.72) is

$$\frac{I(s)}{V(s)} = \frac{1}{s^{\alpha} + 1}, \tag{2.73}$$

which is the fractional equivalent equation of RL circuit. Using the Eq. (2.72), it is convenient to analyse this circuit for different α values which is nothing but different fractional order model of RL circuit. Now, let us chose different α values and check the response of corresponding FO system and compare with the integer order model ($\alpha = 1$). Figure 2.9 depicts the responses of such systems.

For each α value, this RL circuit shows somewhat different behaviour. Hence, utilizing the concept of fractional modeling one can extend the range of modeling from integer model to fractional model which facilitates a more detailed analysis of the considered system. Subsequently, equivalent FO models of robotic systems are derived using similar steps to derive the fractional model as discussed herein.

2.2.1 Embedding to Inverted Pendulum and Cart (POAC)

The state-space model (2.32) and (2.33) represent the POAC system which is considered next for the controller design. Let us find transfer function model from this state space model for $M = 0.5\,\text{Kg}$, $m = 0.2\,\text{Kg}$, $l = 1\,\text{m}$ and $g = 9.8\,\text{m/s}^2$, given as

$$H(s) = \begin{pmatrix} \dfrac{2s^2 - 19.6}{s^4 - 13.72s^2} \\ \dfrac{-2s^2}{s^4 - 13.72s^2} \end{pmatrix}. \tag{2.74}$$

The FO model is considered because FO systems are more flexible when compared to IO systems. There are many papers/books that validate the concept behind using FO models [23]. If we put $\alpha = 1$, then the FO transfer function is the same as the IO transfer function. Using α, the fractional term is embedded to the transfer function using definition of fractional Laplace transform of the POAC system. The fractional equivalent transfer function can be expressed by opting the method used in the papers [24, 25] as

$$H(s) = \begin{pmatrix} \dfrac{2s^{2\alpha} - 19.6}{s^{4\alpha} - 13.72s^{2\alpha}} \\ \dfrac{-2s^{2\alpha}}{s^{4\alpha} - 13.72s^{2\alpha}} \end{pmatrix}. \tag{2.75}$$

Depending on the values of $0 < \alpha \leq 1$, a valid model is derived. For instance, for $\alpha = 0.3$, (2.75) can be approximated using Oustaloup recursive approximation [26] given as

$$\overline{H(s)} = \begin{bmatrix} \dfrac{s + 90}{s^2 + 2s + 1} \\ \dfrac{s + 0.15}{s^2 + 2s + 1} \end{bmatrix}. \tag{2.76}$$

A controller designed for the model (2.76) represents a controller for the FO model of the POAC system with $\alpha = 0.3$. We derive 2nd order approximated systems because the handling of 2nd order system is relatively simpler. Similarly, we can take other values of α to derive the fractional model.

2.2.2 Embedding to 2D-Gantry Crane System

The dynamical equations of 2D-Gantry Crane System are given by (2.48). State-space form represented in (2.50) and (2.51) is used to find the system transfer function given as

$$H(s) = \begin{pmatrix} \dfrac{0.4s^2 + 4}{s^4 + 13.7s^2} \\ \dfrac{-0.4s^2}{s^4 + 13.7s^2} \end{pmatrix}. \tag{2.77}$$

The fractional equivalent transfer function is expressed as

A MATLAB Code on Oustaloup's Approximation of POAC System

Considering $\alpha = 0.3$ in (2.75)

$$H(s) = \begin{pmatrix} \frac{2s^{0.6}-19.6}{s^{1.2}-13.72s^{0.6}} \\ \frac{-2s^{0.6}}{s^{1.2}-13.72s^{0.6}} \end{pmatrix},$$

and writing a MATLAB code to approximate to its equivalent (2.76)

```
clear
clc
w1=.001; w2=1000; N=2;
s=tf('s');

g1=oustafod(0.6 ,N, w1,w2 );
g2=oustafod(0.2 ,N, w1,w2);

G11=(2*g1 -19.6)/(s*g2 -13.72*g1);
G22=(-2*g1)/((s*g2 -13.72*g1));

G1=opt_app(G11 , 1 , 2 , 0);
G2=opt_app(G22 , 1 , 2 , 0);

[num,den] = tfdata(G1,'v');
[num1,den1] = tfdata(G2,'v');

G=tf(num,den);
GG=tf(num1,den1);

H_bar = [G;GG]

_________OUTPUT__________

H_bar =

 From input to output...

        s + 89.99
   1:  ---------------
        s^2 + 2 s + 1

        s + 0.1458
   2:  ---------------
        s^2 + 2 s + 1
```

$$H(s) = \begin{pmatrix} \dfrac{0.4s^{2\alpha} + 4}{s^{4\alpha} + 13.7s^{2\alpha}} \\ \dfrac{-0.4s^{2\alpha}}{s^{4\alpha} + 13.7s^{2\alpha}} \end{pmatrix}. \tag{2.78}$$

By substituting the values of $\alpha = 1$ in (2.78), the corresponding integer order (IO) model in (2.77) is obtained, and in general $0 < \alpha \leq 1$ is a non-integer (fractional) order, that can give more dynamic system information and increases the stability region. The α value selection is critical in fetching an accurate fractional model.

The system is evaluated for various α and is validated by simulation. Depending upon these appropriate α values and for corresponding models, the controller is designed. For example, with $\alpha = 0.3$ a FO model of the 2-D Gantry crane system can be obtained and then with Oustaloup recursive approximation an approximated function can be obtained, such a second order transfer function equivalent to the

FO for $\alpha = 0.3$, is given as

$$\overline{H(s)} = \begin{pmatrix} \dfrac{1.1s + 0.3}{s^2 + 3.6s + 0.01} \\ \dfrac{-1.6s - 0.4}{s^2 + 60.4s + 14.4} \end{pmatrix}. \tag{2.79}$$

For the above approximated FO model (2.79) which represents a FO model of 2-D Gantry crane system, a controller is designed going forward in the subsequent chapters. Likewise, the other fractional models for various values of α can be obtained to check the controller performance.

2.2.3 *Embedding to Missile Launching Vehicle (MLV)*

Now, let us move to a more complex system wherein multi-input and multi-output systems are considered. A missile launch vehicle is that which carries one or more ground-to-ground or ground-to-air missiles, along with the control (personnel) and equipment needed to organize and execute the launch of such weapons [27, 28]. Many of the launch vehicles are manually controlled. For precise launching of missiles, it is must to control the angular movements. Considering state space model (2.65) and (2.66), the transfer function of Model 1 from Table 2.1 is obtained as

$$H(s) = \begin{pmatrix} \dfrac{0.1s^2 + 0.24}{s^4 + 1.64s^2} & \dfrac{-0.03}{s^2 + 1.64} \\ \dfrac{-0.01}{s^2 + 1.64} & \dfrac{-0.004}{s^2 + 1.64} \end{pmatrix}. \tag{2.80}$$

Taking (2.80) and embedding the FO term α to the obtained transfer function, we find the FO transfer function of MLV given as

$$H(s) = \begin{pmatrix} \dfrac{0.1s^{2\alpha} + 0.24}{s^{4\alpha} + 1.64s^{2\alpha}} & \dfrac{-0.03}{s^{2\alpha} + 1.64} \\ \dfrac{-0.01}{s^{2\alpha} + 1.64} & \dfrac{-0.004}{s^{2\alpha} + 1.64} \end{pmatrix}. \tag{2.81}$$

Choosing $0 < \alpha \leq 1$ enables different FO transfer functions. For instance, $\alpha = 0.1$ represents a transfer function given as

$$H(s) = \begin{pmatrix} \dfrac{0.1s^{0.2} + 0.24}{s^{0.4} + 1.64s^{0.2}} & \dfrac{-0.03}{s^{0.2} + 1.64} \\ \dfrac{-0.01}{s^{0.2} + 1.64} & \dfrac{-0.004}{s^{0.2} + 1.64} \end{pmatrix}. \tag{2.82}$$

The approximated second order fractional model of MLV system using the Oustaloup approximation, can be derived as

$$\overline{H(s)} = \begin{pmatrix} \dfrac{1.77s + 0.102}{s^2 + 12.68s + 0.1827} & \dfrac{-0.74s - 0.05}{s^2 + 65.17s + 3} \\ \dfrac{-0.02s - 0.001}{s^2 + 4s + 0.2} & \dfrac{-.003s - 0.003}{s^2 + 2.6s + 1.6} \end{pmatrix}. \tag{2.83}$$

It is expected that a controller designed for such a model would achieve similar desired objectives as the control action on the original IO system.

In this chapter, the basic idea of EL formulation has been discussed and applied to various systems such as MSD system to model and derive the equations of motion. The EL formulation is then applied to some complex systems like POAC, DIPOAC, and MLV. The modeling equations of such systems are then linearized near the equilibrium points. The modeling equations of POAC are derived and will be used for linear control of the cart motion and upward stability of the pendulum subsequently. The modeling equations of DIPOAC are derived and will be used for linear control of the cart motion and upward stability of the two pendulums. The modeling equations of 2D gantry crane system are also derived to control the swing angle of the load and the trolley position. The modeling equations of MLV are derived to control the angular position along y-axis and z-axis. The modeling equations derived in this chapter will be useful in the next chapter where a fractional equivalent control. The definition of Laplace transform for FO systems is applied to determine the fractional model. Fractional equivalent models of POAC, DIPOAC, 2D gantry crane system and MLV system are derived in this chapter, and trail and error steps are used to derive the FO model of the considered systems. To avoid relying on such a method, an algorithm is proposed in Chap. 6 which is one of the interesting contributions of this book. In the next chapter, we present the steps to design Fractional PID controllers of the systems modeled in this chapter.

References

1. Zhang, R., Zou, Q., Cao, Z., Gao, F.: Design of fractional order modeling based extended non-minimal state space MPC for temperature in an industrial electric heating furnace. J. Process Control **56**, 13–22 (2017)
2. Caponetto, R., Fortuna, L., Porto, D.: Parameter tuning of a non integer order PID controller. In: 15th International Symposium on Mathematical Theory of Networks and Systems, Notre Dame, Indiana (2002)
3. Sethi, J., Deb, D., Malakar, M.: Modeling of a wind turbine farm in presence of wake interactions. In: 2011 International Conference On Energy, Automation And Signal (2011). https://doi.org/10.1109/iceas.2011.6147144
4. Liu, Y., Deb, D., Tao, G.: Modeling and multivariable adaptive control of aircraft with synthetic jet actuators. In: 2008 7Th World Congress On Intelligent Control And Automation (2008). https://doi.org/10.1109/wcica.2008.4593263
5. Patel, R., Deb, D., Dey, R., Balas, V.E.: Mathematical Modelling. Intelligent Systems Reference Library 11–28 (2019). https://doi.org/10.1007/978-3-030-18068-3_2
6. Singh, A.P., Kazi, F., Singh, N.M., Vyawahare, V.: Fractional Order controller design for underactuated mechanical systems. In: The 5th IFAC Symposium on Fractional Differentiation and its Applications-FDA (2012)
7. Singh, A.P., Agarwal, H., Srivastava, P.: Fractional order controller design for inverted pendulum on a cart system (POAC). WSEAS Trans. Syst. Control **10**,172–178 (2015)
8. Singh, A.P., Agrawal, H., Srivastava, P.: Robust fractional model predictive controller (FMPC) design for under-actuated robotic systems. Int. J. Control Autom. **11**(7), 2224–2235 (2018)
9. Joshi, S., Singh, A.P.: Comparison of the performance of controllers for under-actuated systems. In: 2016 IEEE 1st International Conference on Power Electronics, Intelligent Control and Energy Systems (ICPEICES). IEEE, pp. 1–5 (2016)
10. Patel, R., Deb, D., Dey, R., Balas, V.E.: Exact linearization of two chamber microbial fuel cell. In: Adaptive and Intelligent Control of Microbial Fuel Cells. Intelligent Systems Reference Library, vol. 161. Springer (2020). https://doi.org/10.1007/978-3-030-18068-3_8
11. Hauser, J., Sastry, S., Kokotovic, P.: Nonlinear control via approximate input-output linearization: the ball and beam example. IEEE Trans. Autom. Control **37**(3), 392–398 (1992)
12. Singh, A.P., Srivastava, T., Srivastava, P.: Fractional equations of motion via fractional modeling of underactuated robotic systems. In: IEEE ICARET, p. 278 (2013)
13. Singh, A.P., Kazi, F.S., Singh, N.M., Srivastava, P.: PI D controller design for underactuated mechanical systems. In: 2012 12th International Conference on control automation robotics & vision (ICARCV). IEEE, pp. 1654–1658 (2012)
14. Singh A.P., Joshi S., Srivastava P.: Predictive controller design to control angle and position of DIPOAC. In: Kulkarni A., Satapathy S., Kang T., Kashan A. (eds.) Proceedings of the 2nd International Conference on Data Engineering and Communication Technology. Advances in Intelligent Systems and Computing, vol 828. Springer, Singapore (2019)
15. Singh, Abhaya Pal., Agrawal, Himanshu., Srivastava, Pallavi, Naidu, Praveen V.: A robust fractional model predictive control (FMPC) design. Progr. Fract. Differ. Appl. **5**(3), 217–223 (2019)
16. Singh, A., Agrawal, H.: A fractional model predictive control design for 2-d gantry crane system. J. Eng. Sci. Technol. **13**(7), 2224–2235 (2018)
17. Singh, A.P., Srivastava, T., Agrawal, H., Srivastava, P.: Fractional order controller design and analysis for crane system. Progress Fract. Differ. Appl. Int. J. **3**(2), 155–162 (2017)
18. Singh, A.P., Agrawal, H.: On modeling and analysis of launch vehicle system. In: Proceedings of the International Conference on Signal, Networks, Computing, and Systems. Springer, New Delhi, pp. 273–279 (2016)
19. Caponetto, R.: A new tuning strategy for a non-integer order PID controller. In: First IFAC workshop on fractional differentiation and its application, Bordeaux, France, pp. 168–173 (2004)

20. Kexue, L., Jigen, P.: Laplace transform and fractional differential equations. Appl. Math. Lett. **24**(12), 2019–2023 (2011)
21. Singh, A.P., Deb, D., Agarwal, H.: On selection of improved fractional model and control of different systems with experimental validation. Commun. Nonlinear Sci. Numer. Simul. **79**, 104902 (2019)
22. Shah, P., Agashe, S.D., Singh, A.P.: Fractional order modelling using state space theory. Int. J. Eng. Technol. **5**(3), 2891–2894 (2013)
23. Ahmed, E., El-Sayed, A.M.A., El-Saka, H.A.: Equilibrium points, stability and numerical solutions of fractional-order predator-prey and rabies models. J. Math. Anal. Appl. **325**(1), 542–553 (2007)
24. Ahmad, W.M., El-Khazali, R.: Fractional-order dynamical models of love. Chaos Solitons Fractals **33**(4), 1367–1375 (2007)
25. Song, L., Xu, S., Yang, J.: Dynamical models of happiness with fractional order. Commun. Nonlinear Sci. Numer. Simul. **15**(3), 616–628 (2010)
26. Xue, D., Zhao, C., Chen, Y.: A modified approximation method of fractional order system. In: 2006 International Conference on Mechatronics and Automation. IEEE, pp. 1043–1048 (2006)
27. Viswanath, D., Deb, D.: A new nonlinear guidance law formulation for proportional navigation guidance. In: 2012 12Th International Workshop On Variable Structure Systems (2012). https://doi.org/10.1109/vss.2012.6163500
28. Viswanath, D., Krishnaswamy, S., Deb, D.: Homing missile guidance using LOS rate and relative range measurement. In: 2015 Annual IEEE India Conference (INDICON) (2015). https://doi.org/10.1109/indicon.2015.7443715

Chapter 3
FOPID Controller Design for IO Model of Robotic Systems

The last chapter discussed the modeling of different type of robotic systems. The overall system output performance depends on system model accuracy, and the problem worsens if an inaccurate system model is chosen [1, 2]. Once the model of a system is selected, an efficient controller is needed to control the output response of the system [3–5]. Designing controller for a given system is another major issue, and FO controllers have been demonstrated to be highly effective even in this field. Podlubny introduced the term FOPID, and designed the controller with a FO integrator and a FO differentiator for FO systems [6, 7]. FOPID controllers utilize five tuning parameters compared to Integer Order Controller which have only three tuning parameters, and hence are able to provide more flexibility to implement the controller precisely and more accurately to fit the needs for a system [8, 9]. In this chapter, we discuss how to design an Integer and Fractional PID controllers for various robotic systems. The response of such systems is also compared to check the effectiveness of the designed controller.

A good controller design provides a better system output by incorporating a controlled change at the input applied to a system. It is easy to pic PID controller which proved to be efficient, robust and most popular control system in designing a controller. PID controller is utilized to obtain a robust output from a system. PID controller is one of the most widely used control techniques in process control, robotics and many other fields where set-point needs to be maintained. It is used to minimize or nullify the error between set-point and the measured variable.

For instance, let's say that one has a small toy bike that can move forward and backwards. Assume that one can tell the bike where it ought to be (the desired location) and the bike can know where its actual location is. Let us consider that we want to make the bike move quickly and accurately to a new location. One can apply electricity to the motor of the bike but one does not know how much electricity to apply. The solution to this problem is PID control. The difference between the desired location (DL) and the current location (CL) is something we call as the error, $E = DL - CL$. For PID control there are three different terms known as gain factors: K_p, K_d and K_i that we can mathematically manipulate with the error to check

A. P. Singh et al., *Fractional Modeling and Controller Design of Robotic Manipulators*, Intelligent Systems Reference Library 194,
https://doi.org/10.1007/978-3-030-58247-0_3

exactly how much electricity input should we apply to the motor at any point of time. The proportional gain (K_p) means that the electricity input applies to the motor is in proportion to the error. A large error means large electricity input to correct it. If the proportional gain is too low then the bike will just move slowly. If one keep increasing the proportional gain, it moves faster but eventually, the bike moves past the desired location and then runs back and forth several times eventually stopping near the desired location. This is why we need a derivative gain (K_d). Derivative gain means that if there is error but going down quickly (the bike is moving fast), then one can decrease the electricity applied to the motor by a small portion. This prevents moving past the desired position and helps to stop at or near to the desired position directly. But even with the proportional gain K_p and derivative gain K_d, the bike will not land exactly at the desired position because even if one is close, such a small error will not make enough electricity input to move the car at all. To solve this problem we keep track of the accumulated error over time and use that to add some electricity so that the car does not stop short of the target. This is called integral gain K_i and ensures that one moves to the desired position. The best values for K_p, K_d and K_i can be found by some mathematical modeling or by trial and error. The best part is that after one finds the ideal values for all 3 gains, the bike will move with optimum efficiency and performance for all conditions.

Subsequently, a fractional order PID (FOPID) control have been developed [12–15]. These types of PID controllers have two more manipulating variables, one each of integral and derivative orders. A total of five different control specifications can be found when using a FOPID control strategy and this is two more when compared to the traditional PID control. Because of these, FOPID controllers gives better results compared to the traditional PID controllers. Fractional PID control is further studied to design controllers for different robotic systems.

3.1 Controller Design for Mass-Spring System

Let us consider a mass-spring system as shown in Fig. 3.1 to design the traditional PID controller and FOPID controller to make this system reach a desired target. The modeling equation can be found by following the steps mentioned in Chap. 2, while considering the frictional force B, and is given as

$$M\ddot{x} + B\dot{x} + kx = F. \tag{3.1}$$

Taking the Laplace transform, we get the transfer function

$$\frac{X(s)}{F(s)} = \frac{1}{Ms^2 + Bs + k}. \tag{3.2}$$

Let us use (3.2) for designing the controller. A control system block diagram is easy to follow corresponding to (3.2) and is shown in Fig. 3.2.

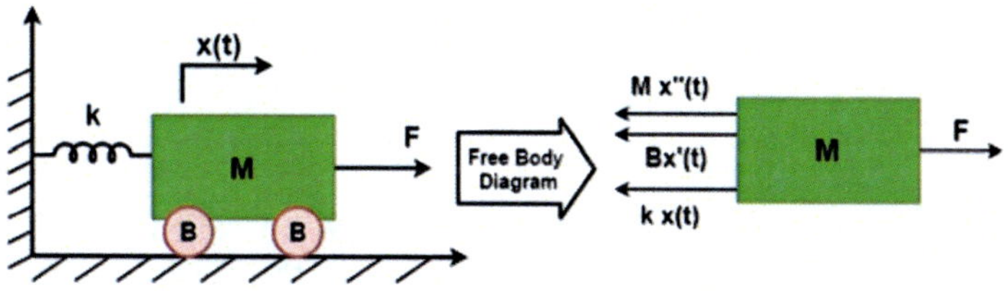

Fig. 3.1 Mass-spring system

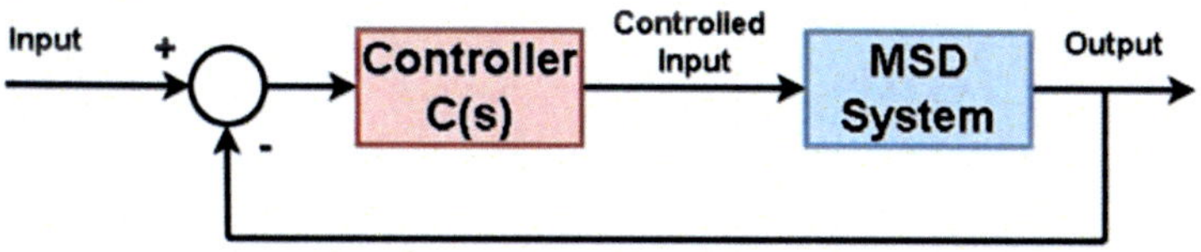

Fig. 3.2 Control system block diagram of MSD system

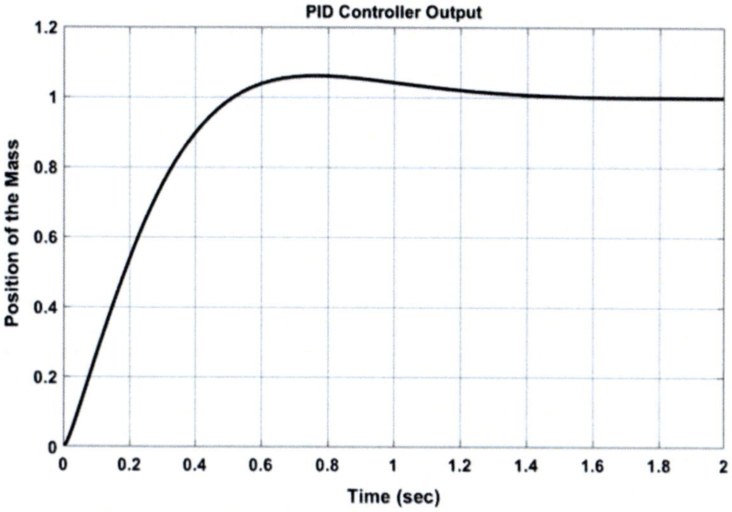

Fig. 3.3 PID controller response to MSD system

Now, let us design a PID controller which can provide a desired output. Such a transfer function is

$$C(s) = k_p + \frac{k_i}{s} + sk_d \tag{3.3}$$

Choosing the values of $K_p = 39$, $K_d = 2.4$ and $K_i = 115$, and assuming that we want to stop the system at position $x(t) = 1m$ within 2 s of applying a constant force on the mass and assuming $m = 1$ kg, $b = 10$ Ns/m, $k = 20$ N/m and $F = 1$ N. The PID controller response is shown in Fig. 3.3.

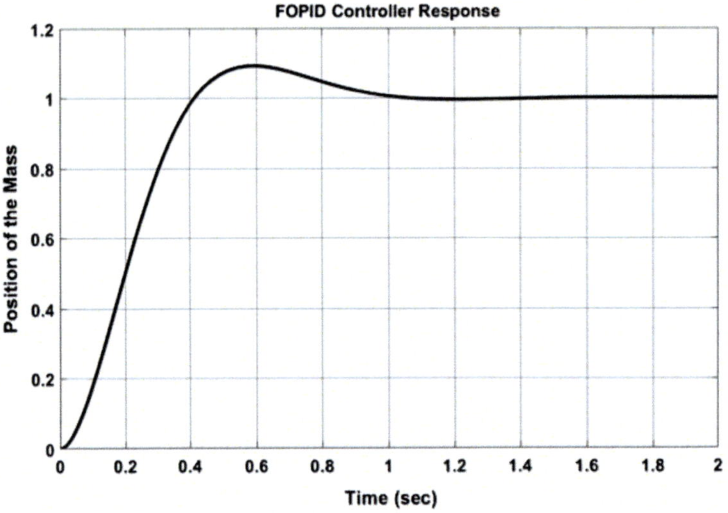

Fig. 3.4 FOPID controller response to MSD system

Now, let us choose the same controller gain values of $K_p = 39$, $K_d = 2.4$ and $K_i = 115$ in designing the FOPID controller

$$C(s) = k_p + \frac{k_i}{s^{\alpha}} + s^{\beta} k_d. \tag{3.4}$$

The response of this FOPID controller is shown in Fig. 3.4.

Comparing the results obtained from PID and FOPID (Fig. 3.5), it is concluded that the FOPID response settles faster compared to its counterpart PID. In the FOPID controller, the values of $\alpha = 0.9$ and $\beta = 0.4$ are chosen.

3.2 PID and Fractional PID Controller Design

Next, we discuss two cases of FO controller design. Firstly, the controller is designed without considering the system damping, and then the system damping is considered to design a FOPID + Damping controller. The transfer function of Integer PID controller can be written as

$$k_{pid} = k_p + \frac{k_i}{s} + k_d s, \tag{3.5}$$

and FO PID controller can be written as

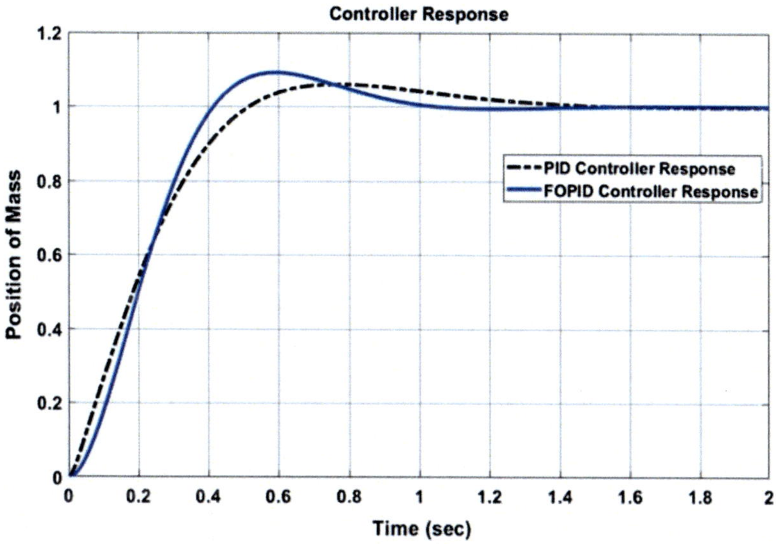

Fig. 3.5 Controller response comparison to MSD system

$$k_{fopid} = k_p + \frac{k_i}{s^{\alpha}} + k_d s^{\beta}, \tag{3.6}$$

where $0 < \alpha < 1$, and $0 < \beta < 1$.

3.2.1 For POAC System

Three cases of controller designs are presented here and the responses are compared:

(i) Fractional PID with Damping (FOPID+Damping) controller
(ii) Fractional PID (FOPID) controller
(iii) Traditional PID controller (IOPID)

Figure 3.6 presents the Simulink model of the whole controlled POAC system. SET block consists of the desired points required to make the system stable in minimum time. An external force is put on the cart and a controller is designed to control the system stability. For simulation, these parameters are considered: $m = 0.2\,\text{Kg}$, $M = 0.5\,\text{Kg}, l = 1\,\text{m}, g = 9.8\,\text{m/s}^2$ and $B = 1\,\text{Kg/s}$. A classical order PID controller can be obtained for $\alpha = \beta = 1$ in (3.6). Also with $\alpha = 1$, $\beta = 0$ and $\alpha = 0$, $\beta = 1$ in (3.6) and comparing it with the traditional PD and PI controllers, it is clear that the classical PID controller is a subset of the fractional $PI^{\alpha}D^{\beta}$ controller. The fractional controller is a generalization of the classical. The $PI^{\alpha}D^{\beta}$ controller should improve the transient system performance. FOPID controller utilizes five tuning parameters, i.e. $(K_d, K_p, K_i, \beta, \alpha)$, and hence provides more flexibility.

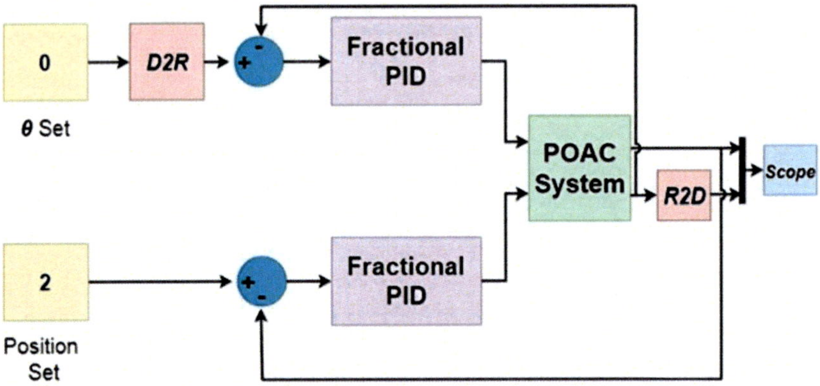

Fig. 3.6 Fractional order controller simulink model

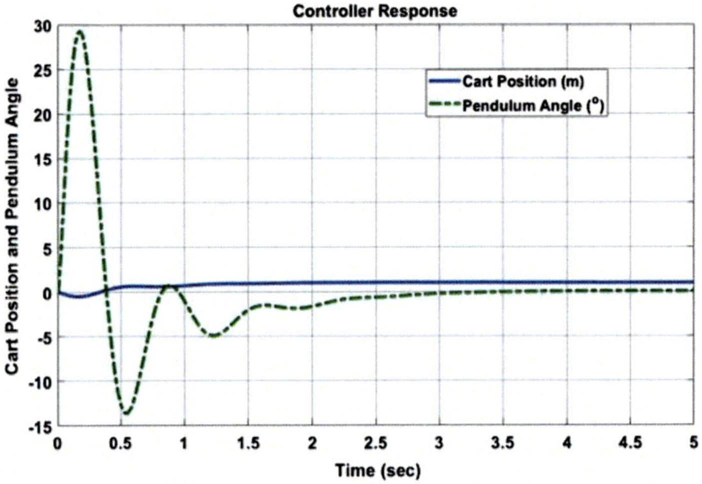

Fig. 3.7 Fractional order controlled response

A FOPID controller is designed for an IO system whose state-space model is already obtained in (2.32) in Chap. 2, and the performance is compared with the traditional PID controller. We use MATLAB Simulink to analyze the performances of both the controllers. From Figs. 3.7 and 3.8, we identify that the FOPID controller designed for POAC system minimises the value of overshoot and settling time when compared to the counterpart IOPID controller. The values of tuned controlled parameters are provided in Tables 3.1 and 3.2, and using these, a FOPID controller is obtained.

The controller is designed such that the position of cart is stable at desired position with vertically upward pendulum i.e., pendulum angle to zero. Figure 3.7 shows the output performance of FOPID controller, Fig. 3.8 shows the output performance of

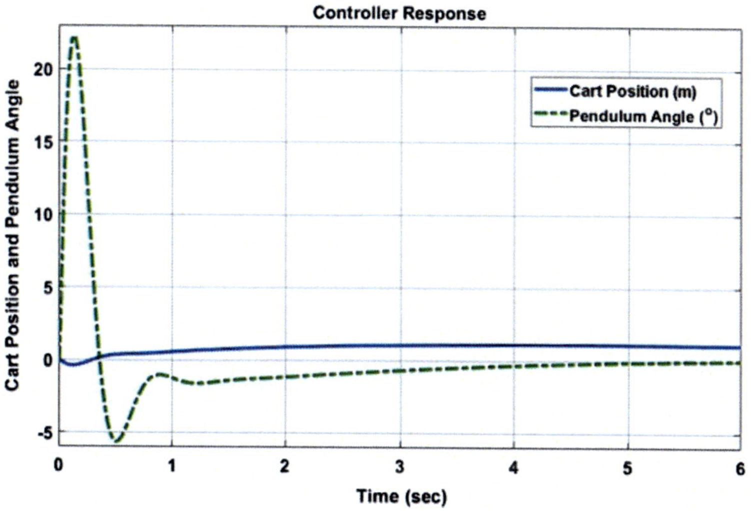

Fig. 3.8 Integer order controlled response

Table 3.1 Optimal values of parameters (angle control of pendulum)

Controller	α	β	k_p	k_d	k_i
PID	1	1	−55	−8	−10
FOPID	0.95	0.9	−55	−8	−25
FOPID+Damping	0.95	0.9	−60	−8	−25

Table 3.2 Optimal values of parameters (position control of cart)

Controller	α	β	k_p	k_d	k_i
PID	1	1	−1.7416	−3.192	−0.0128
FOPID	0.009	0.8	−1.8	−5	−0.8
FOPID+Damping	0.009	0.8	−1.7416	−6	−0.0128

IOPID controller, and Fig. 3.9 shows the output performance of FOPID controller with damping of the system. It is observed that the controllers which is designed using fractional calculus concept improves the system performance significantly.

Figure 3.10 shows the output performance comparison of the three controllers which are designed to control the position of the cart. It can be noticed that the time to settle the response of IOPID controller to make the cart position stable is 10 sec, which is improved to about 5 sec when using the FOPID controller, and about 4 sec with damping and a fractional controller. Figure 3.11 shows the output response comparison of the three controllers which are designed to control pendulum angle.

The time to settle with IOPID controller to make the cart position stable is 8 sec which can be improved using the FOPID controller (5.5 sec), and further improved

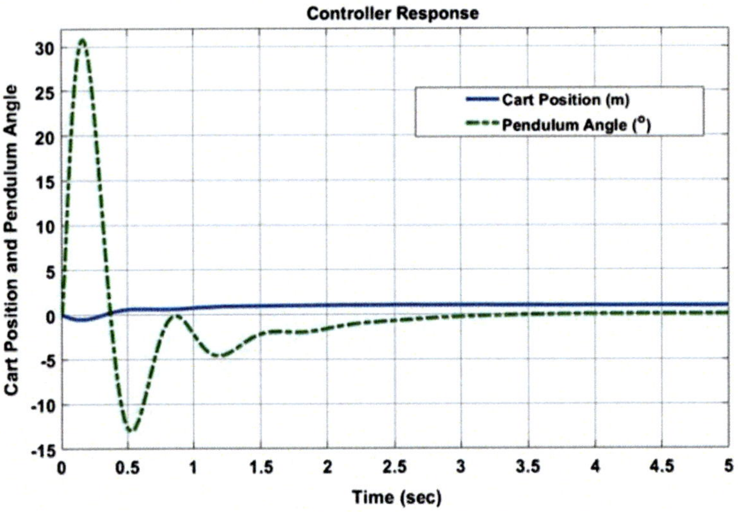

Fig. 3.9 Fractional order controlled response with damping

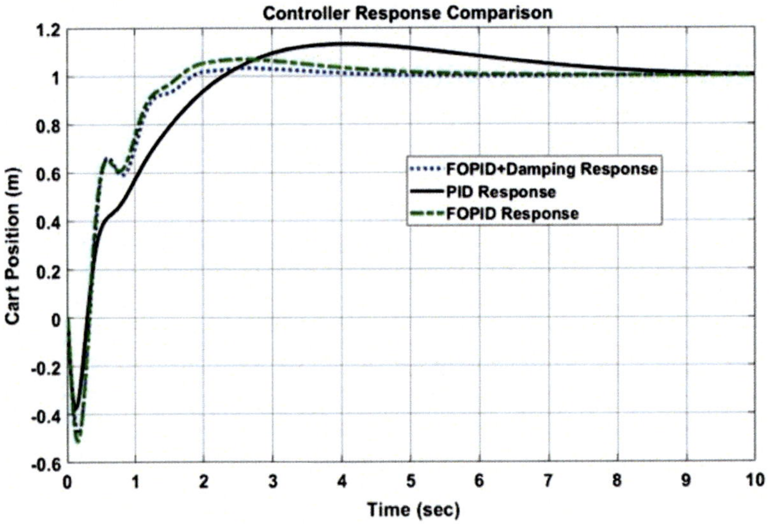

Fig. 3.10 Cart position control comparison

with damping factor along with a fractional controller (3.5 sec). Simulations indicate that when FOPID controller is designed, the system performance is improved compared to classical IOPID controller, and FOPID controller designed considering system damping, gives even better outcome.

The FOPID controller for POAC system is designed and the performance is compared with the traditional PID. The FOPID controller stabilizes the system in mini-

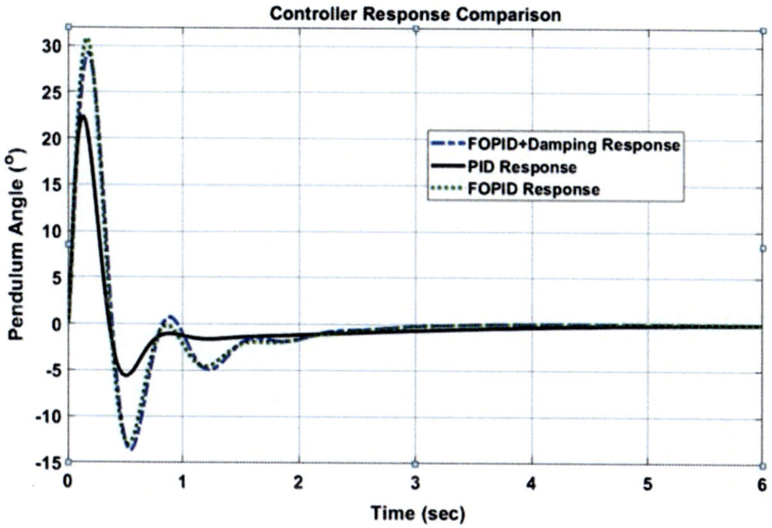

Fig. 3.11 Pendulum angle control comparison

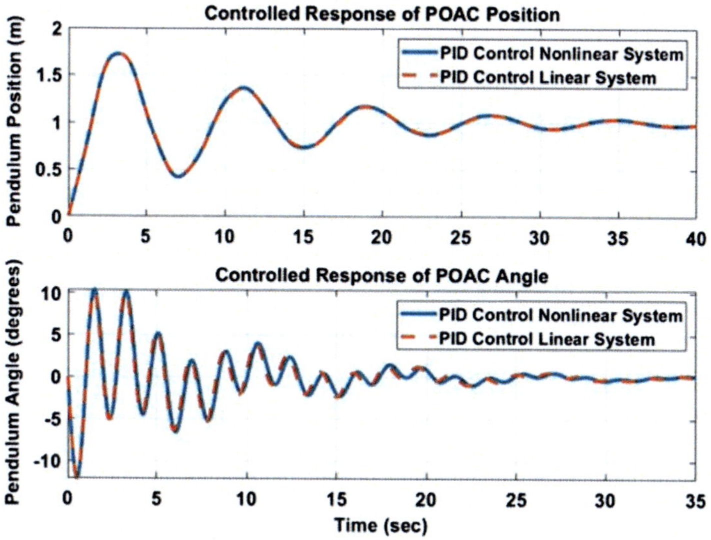

Fig. 3.12 PID controller response of POAC system

mum time. The transient response in Fig. 3.11 is not good and the angle overshoots to around 31°. From Fig. 3.12, it can be observed that the linear and the nonlinear systems respond almost identically to the PID controller. POAC position takes 50+ s to be stable and the angle takes 35+ s with limited overshoot.

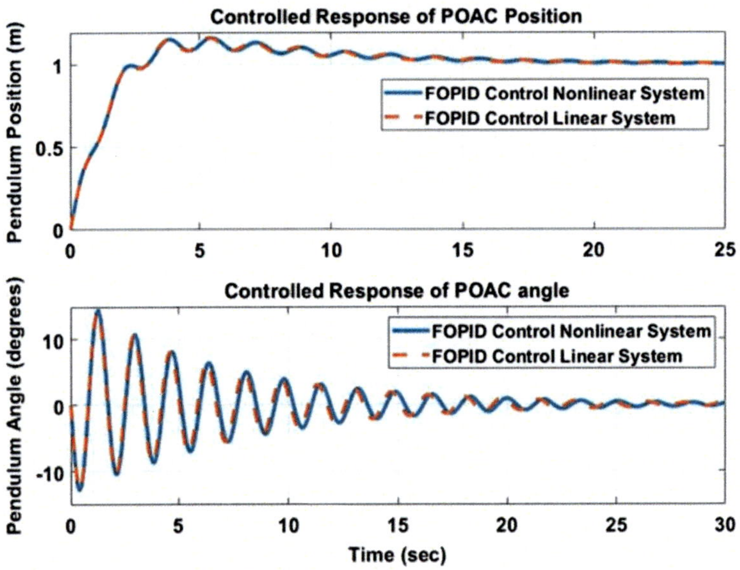

Fig. 3.13 FOPID controller response of POAC system

From Fig. 3.13, it can be observed that the linear and nonlinear systems respond identically to the FOPID controller. POAC Position takes 20 s to be stable and the angle takes 30 s to be stable within the limited overshoot.

Figure 3.14 shows the FOPID and PID controller response comparison. It can be observed that the FOPID controller designed to control the position and angle of the system takes less time compared to the IOPID controller responses for both linear and nonlinear system model of POAC.

3.2.2 For 2D Gantry Crane System

Next, we design the FOPID controller for 2D Gantry Crane system and evaluate the system performance. For comparison, the IOPID and FOPID controllers are designed to find the system responses. Two controllers are present in the system which controls the system performance in parallel. The first controller controls the position of the trolley and the second controller controls the swing angle. The parameter values can be obtained from an experimental set up discussed in [10] of gantry crane system. Considering those system parameters, a controller is designed to control the position and angle of this system. The following desired conditions are mentioned to accomplish faster response from the system:

(i) Settling time $\prec 7s$, (ii) Steady state error $\leq 1\%$, (iii) Overshoot $\leq 22.5\%$.

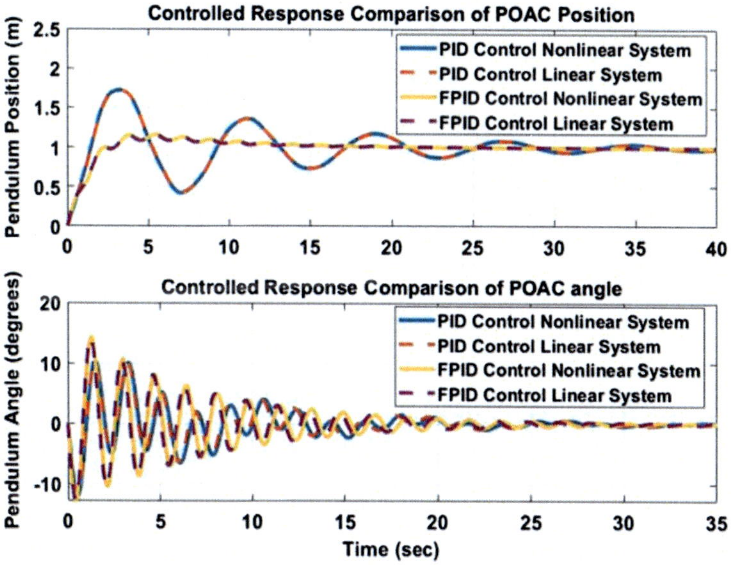

Fig. 3.14 Controller response comparison of POAC system

Table 3.3 FOPID controller parameters

Controllers	k_p	k_i	k_d	α	β
FOPID angle controller	−1	4	−5	0.8	0.99
FOPID position controller	7	2	7	0.9	0.8

The IOPID and FOPID controllers are designed with the same values for k_p, k_i and k_d, to study the effectiveness of FOPID which has two more parameters to tune. The α and β values are chosen in between 0 and 1. The inbuilt PID tuner block is used to find the PID and FOPID parameter values as shown in Table 3.3.

With FOPID controller, the swing angle is controlled in less time. From Figs. 3.15 and 3.16, it is noticed that the settling time for the control of θ as well as position with FOPID controller is less compared to IOPID controller. The settling times for the responses with the two controllers are given in Table 3.4.

Results show that FOPID proves a better controller and controls the response faster as compared to traditional PID. The FOPID controllers have two additional parameters to tune which provide more flexibility. The simulation results show that the FOPID controller performance is better compared to the IOPID controllers when controlling the swing angle as well as the position of 2-D Gantry crane system.

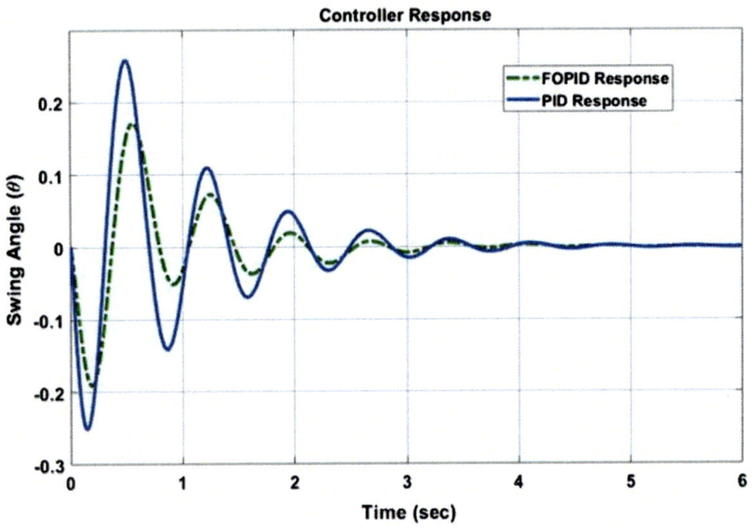

Fig. 3.15 Angle comparison of the controllers

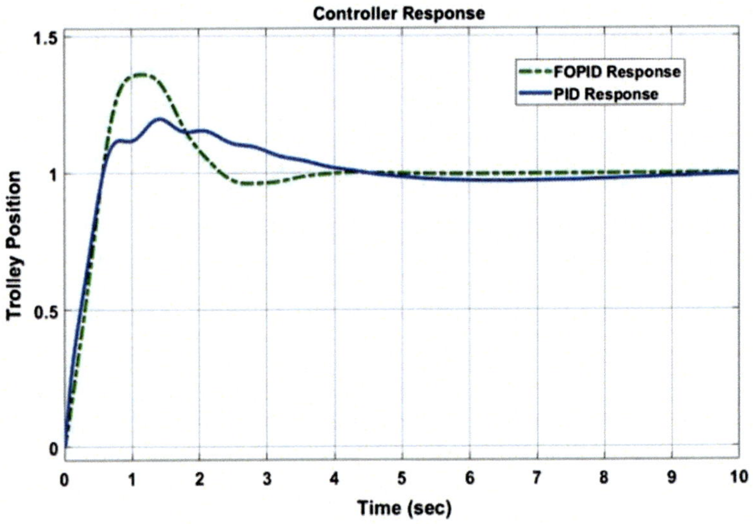

Fig. 3.16 Position comparison of the controllers

3.3 Model Predictive Controller (MPC)

Next, we discuss two cases of controller design. In one case, the system damping is not considered for controller design and in the other, the system damping is considered

Table 3.4 Comparison of controllers

Performance	PID	FOPID
Settling time in Sec (Angle)	3.5	5.5
Settling time in Sec (Position)	4	10

Table 3.5 PID parameters

Controllers	k_p	k_i	k_d
PID angle controller	−55	−8	−10
PID position controller	−1.74	−3.19	−0.013

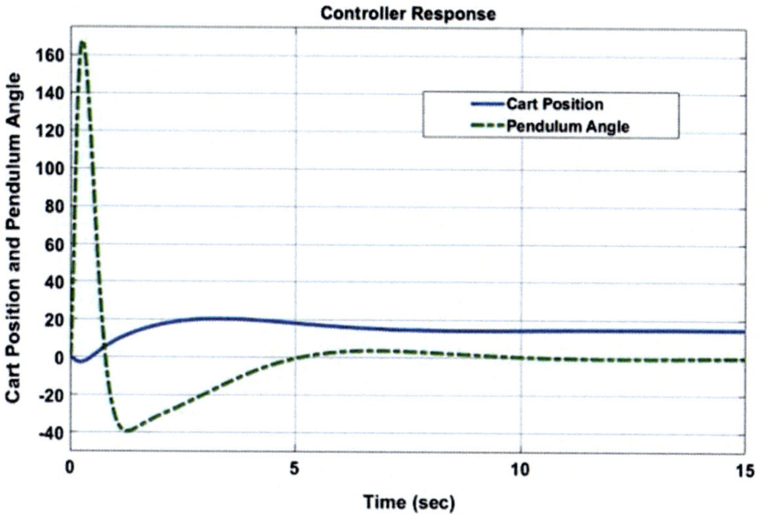

Fig. 3.17 PID response at cart's position 15

to design a controller using the MPC toolbox available in MATLAB [11]. The ideal estimations of PID controller parameters are shown in Table 3.5.

PID controllers needs the tuning parameters to optimize the response. Prediction Horizon, Control Horizon, and Control Interval are the three parameters of MPC which need to be set for a desired output. Figure 3.17 shows the PID controller output response when the cart distance from the origin is 15 m.

Figure 3.18 demonstrates the basic MPC structure consisting of inputs and outputs with disturbances. POAC system has two degrees of freedom and is under-actuated. The system consists of one input and two outputs, one of which is measured and the other is not measured. The measured output is then fed back and accordingly the output of MPC is obtained. The system parameters are $m = 0.2\,\text{kg}$, $M = 0.5\,\text{kg}$, $L = 1\,\text{m}$, $G = 9.8\,\text{m/s}^2$, $B = l\,\text{kg/s}$. For MPC, the Control interval is set to 0.1 sec, Prediction horizon to 30 and Control horizon to 6.

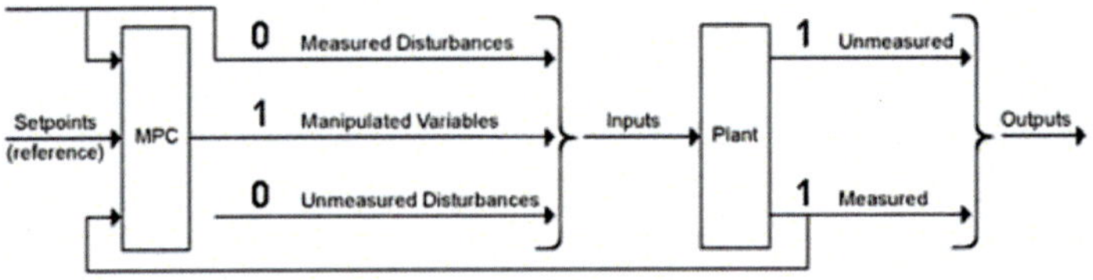

Fig. 3.18 MPC block diagram

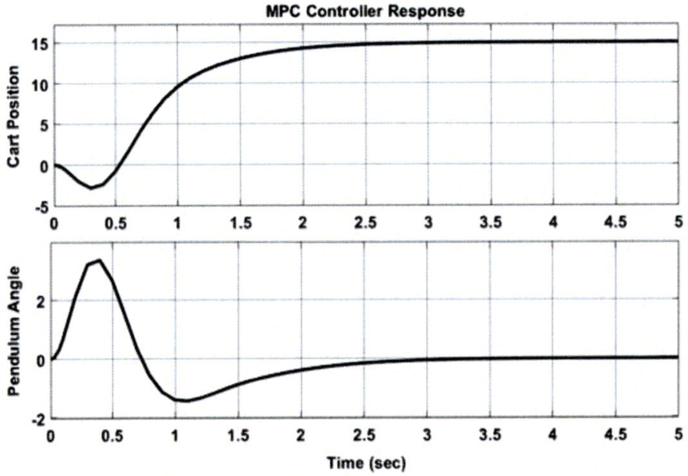

Fig. 3.19 MPC response (output) at carts position 15 without damping

For the POAC system, the MPC controller is expected to settle the position and the vertically upward angle to 15 m and zero degrees respectively in the least amount of time with less overshoot. Figure 3.15 shows the output response of the MPC controller when the damping of POAC system is not considered. It can be noticed from Figs. 3.17 and 3.19 that the MPC is a good option to control the system performance and takes less amount of time to settle with little overshoot (Fig. 3.20).

Figure 3.21 shows the output response of the MPC controller with POAC system damping. It is noticed from Figs. 3.21 and 3.17 that the MPC with system damping is efficient and takes lesser amount of time to settle with little overshoot. Figure 3.22 shows the input that is applied to the controlled system to get stable response in relatively less time. The simulation shows that if the MPC method is used to design the controller of POAC system then the system response settles faster compared to a PID controller for the same system. Hence, MPC is the better option for the POAC system. Table 3.6 shows the output response comparison of these controllers. It can be easily noticed from the table that MPC performs better compared to PID.

An idea of how to represent a transfer function of the fractional PID controller is discussed and a method of controller design is presented to control the position of the cart and angle of the pendulum. The fractional controller designed for the

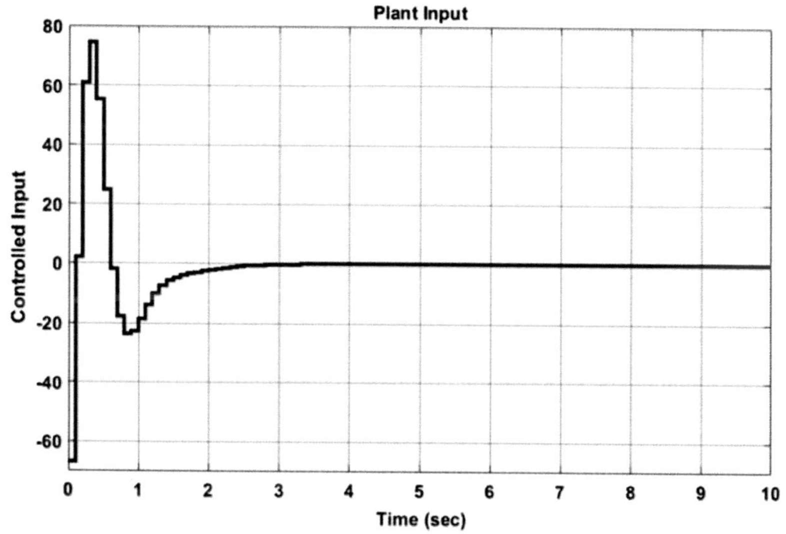

Fig. 3.20 MPC response (input) at cart's position 15 without damping

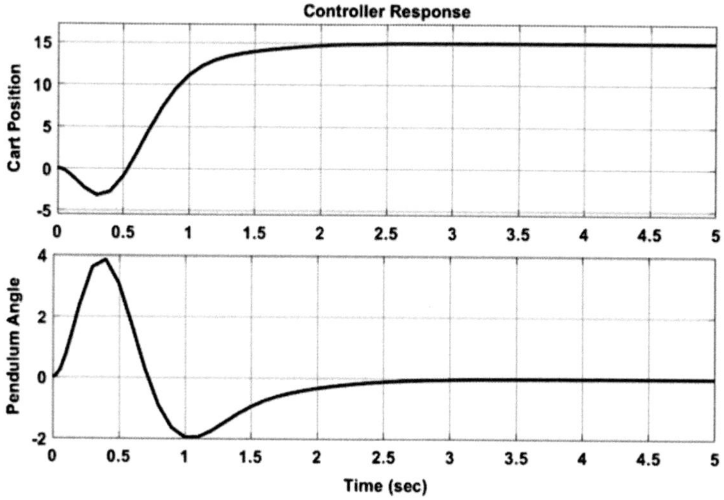

Fig. 3.21 MPC response (outputs) at cart's position 15 with damping

POAC system performs better compared to traditional PID controller response. A method of designing a fractional controller for 2D gantry crane system is presented and the results with different controllers are summarized. The fractional controller performs better in terms of settling time. A model predictive controller design is also presented for POAC system and the output responses are compared with the

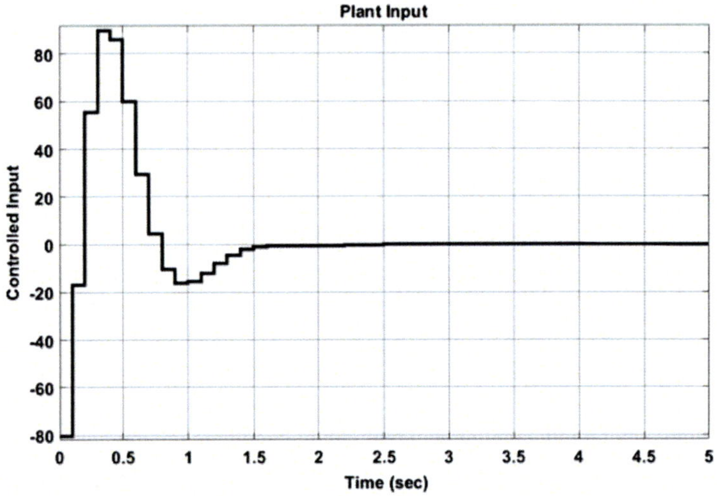

Fig. 3.22 MPC response (input) at cart's position 15 with damping

Table 3.6 Comparison table

Controllers and specifications	PID	MPC	MPC + Damping
Settling time of angle controller	10 Sec	3 Sec	2.5 Sec
Settling time of position controller	10 Sec	2.8 Sec	2.5 Sec
Overshoot in angle and position	Very high	Less	Less

PID control method and it is found that with MPC method, hen the system performs better. A Fractional MPC controller design method is discussed in the next chapter.

References

1. Liu, Y., Deb, D., Tao, G.: Modeling and multivariable adaptive control of aircraft with synthetic jet actuators. In: 2008 7th World Congress On Intelligent Control And Automation (2008). https://doi.org/10.1109/wcica.2008.4593263
2. Deb, D., Tao, G., Burkholder, J.: High-order design of adaptive inverses for signal-dependent actuator nonlinearities. In: 2008 American Control Conference (2008)
3. Patel, R., Deb, D., Modi, H., Shah, S.: Adaptive backstepping control scheme with integral action for quanser 2-dof helicopter. In: 2017 International Conference On Advances In Computing, Communications And Informatics (ICACCI) (2017). https://doi.org/10.1109/icacci.2017.8125901

4. Rajkumar, S., Chakraborty, S., Dey, R., Deb, D.: Online delay estimation and adaptive compensation in wireless networked system: an embedded control design. In: International Journal Of Control, Automation And Systems (2019)
5. Malepati, B., Vijay, D., Deb, D., Manjunath, K.: Parameter validation to ascertain voltage of Li-ion battery through adaptive control. Adv. Intell. Syst. Comput. **21–31**, (2018). https://doi.org/10.1007/978-981-13-1966-2_3
6. Podlubny, I.: Fractional-order systems and PI^{λ} D^{μ}-controllers. IEEE Trans. Autom. Control **44**(1), 208–214 (1999)
7. Podlubny, I.: Fractional-order systems and fractional-order controllers. Inst. Exp. Phys. Slovak Acad. Sci. Kosice **12**(3), 1–18 (1994)
8. Singh, A.P., Kazi, F., Singh, N.M., Vyawahare, V.: Fractional Order controller design for underactuated mechanical systems. In: The 5th IFAC Symposium on Fractional Differentiation and its Applications-FDA (2012)
9. Singh, A.P., Deb, D., Agarwal, H.: On selection of improved fractional model and control of different systems with experimental validation. Commun. Nonlinear Sci. Numer. Simul. **79**, 104902 (2019)
10. Wahyudi, M.I.S.: Sensorless anti-swing control for automatic gantry crane system: model-based approach. Int. J. Appl. Eng. Res. **2**(1), 147–161 (2007)
11. Bemporad, A., Morari, M., Ricker, N.L.: Model predictive control toolbox user's guide. The mathworks (2010)
12. Singh, A.P., Srivastava, T., Agrawal, H., Srivastava, P.: Fractional order controller design and analysis for crane system. Progress Fract. Differ. Appl. Int. J. **3**(2), 155–162 (2017)
13. Singh, A.P., Agarwal, H., Srivastava, P.: Fractional order controller design for inverted pendulum on a cart system (POAC). WSEAS Trans. Syst. Control **10**, 172–178 (2015)
14. Shah, P., Agashe, S.D., Singh, A.P.: Design of fractional order controller for undamped control system. In: 2013 Nirma University international conference on engineering (NUiCONE). IEEE, pp. 1–5 (2013)
15. Singh, A.P., Kazi, F.S., Singh, N.M., Srivastava, P.: PI D controller design for underactuated mechanical systems. In: 2012 12th International Conference on cOntrol Automation Robotics & Vision (ICARCV). IEEE, pp. 1654–1658 (2012)

Chapter 4
Fractional Model Predictive and Adaptive Fractional Model Predictive Controller Design

Recently, model predictive controller design for the FO model of dynamical systems has attracted many researchers. In [1], authors discuss how to design generalized predictive control for a fuel cell using fractional calculus. In [2], authors discuss the designing of switched state MPC for fractional-order discrete-time systems. In paper [4], discrete-time FO systems have been considered for MPC controller design. Articles [3] and [5] uses the same methodology to design control law to control the response of the system efficiently. More recently, paper [6] discuss developing a fractional MPC which can be used to control the temperature efficiently in industrial processes. In [3], an idea of how to design MPC for FO systems is presented which is built upon in this chapter to obtain the controller design for a different system. In this chapter, a FO model of a system is chosen using the method discussed in Chap. 3 and then the MPC controller is obtained for efficient control of the considered system. The fractional MPC (FMPC) control design method has tremendous potential to improve system performance. However, some properties of FMPC controller design is at present a subject of in-depth analysis and experimentation.

In the last chapter, a FO control design method is presented to control the IO model of robotic systems. This chapter gives a clear idea about designing FMPC for various robotic manipulators. MPC methods are used to control different systems, i.e., Pendulum on a cart system, 2D Gantry crane system and missile launching vehicle/pad (MLV) system. The aim here is to obtain MPC control using fractional control point of view. MPC toolbox of MATLAB is considered in designing the controller. Simulation experiment reveals the effectiveness of the controller and improved control performance when compared to traditional MPC.

A. P. Singh et al., *Fractional Modeling and Controller Design of Robotic Manipulators*, Intelligent Systems Reference Library 194,
https://doi.org/10.1007/978-3-030-58247-0_4

4.1 Fractional Model Predictive Controller (FMPC) Design for POAC System

In this section, a novel robust FMPC is designed for an under-actuated POAC robotic system which has two outputs: linear position and angular position (i.e., 2 degrees of freedom) and one control input. The fractional equivalent model of the system is obtained and approximated using the Oustaloup recursive approximation. From the approximated model the best model is chosen, and FMPC is designed [7–9]. The FMPC is robust to the variations of the system parameters for an inverted pendulum on a cart system. This is the first instance where FMPC is implemented on the POAC system. For controller design, MATLAB MPC toolbox [10] is used. Figure 4.1 represents the MPC toolbox structure wherein it can be observed that there is one input which is applied on the cart and two outputs: (i) the cart's position is measured and the (ii) pendulum's angle, is not measured. The objective is to control the angle of the pendulum and the cart's position.

Let us consider five different cases of the models depending on α values for controller design to understand the controlled behaviour of the POAC system. These values are arbitrarily chosen as $\alpha = 0.2, 0.3, 0.5, 0.8$ and 1. For $\alpha = 0.3$, let the parameters for MPC design be control horizon as 6, prediction horizon as 30 and sampling time as 0.1 sec. The controlled response is shown in Fig. 4.2.

It is noticed that the MPC controller designed for the FO model of POAC controls the output response in minimum time with minimum overshoot. These results are compared with the those obtained in the previous chapter and the system performance is found to improve with the FMPC controller for the FO model of POAC. Figure 4.3 shows the controlled output response and controlled input response of the MPC controller applied to the FO model of POAC system. The controller design for $\alpha =$ 0.3 performs better in terms of settling time for both the position and the angle. So the fractional model of POAC corresponding to $\alpha = 0.3$ is used to design the FMPC which gives stable and controlled response.

Next, let us consider other values of α representing the original system model, and we apply the same condition considered in the $\alpha = 0.3$ case to design the MPC controller. The response corresponds to the traditional IO MPC model of the POAC system shown in Figs. 4.4 and 4.5. The responses of Fig. 4.2 when compared with Fig. 4.4 shows that FMPC performs better in terms of settling time.

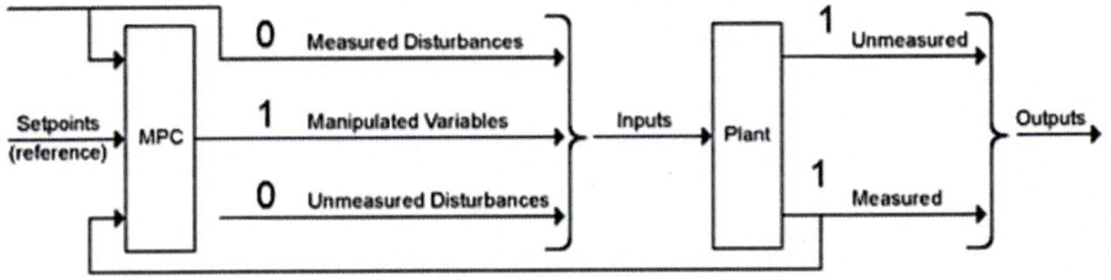

Fig. 4.1 MPC structure

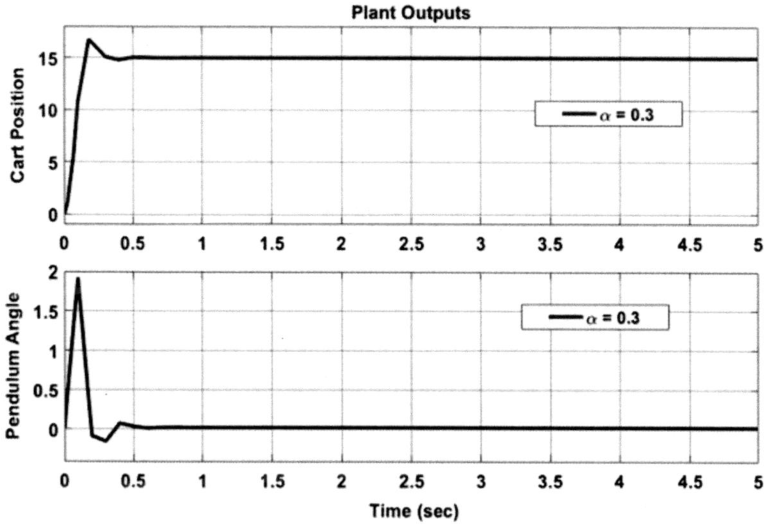

Fig. 4.2 Controlled outputs for $\alpha = 0.3$

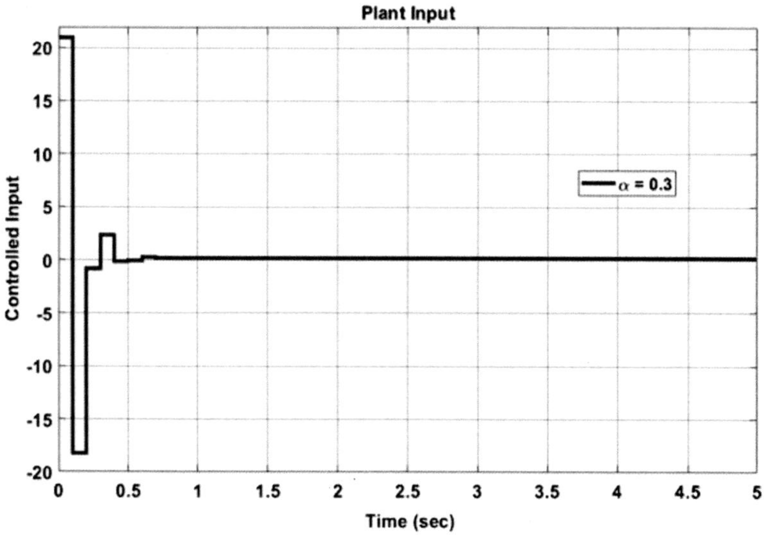

Fig. 4.3 Controlled input for $\alpha = 0.3$

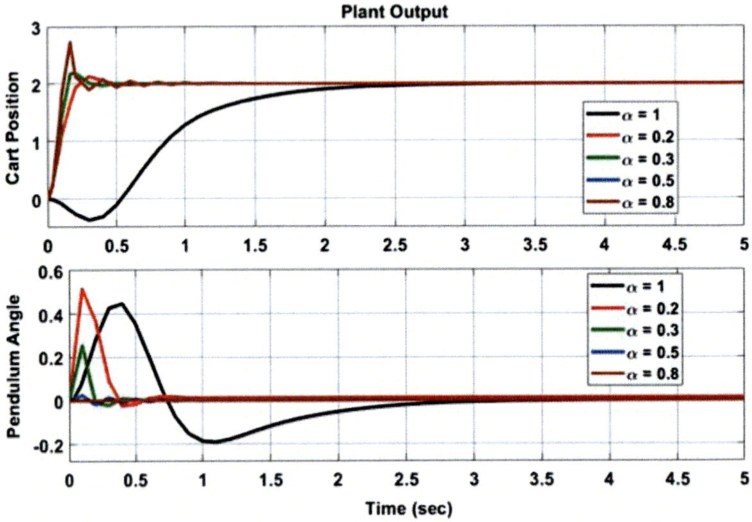

Fig. 4.4 Controlled output for $\alpha = 0.2, 0.3, 0.5, 0.8, 1$

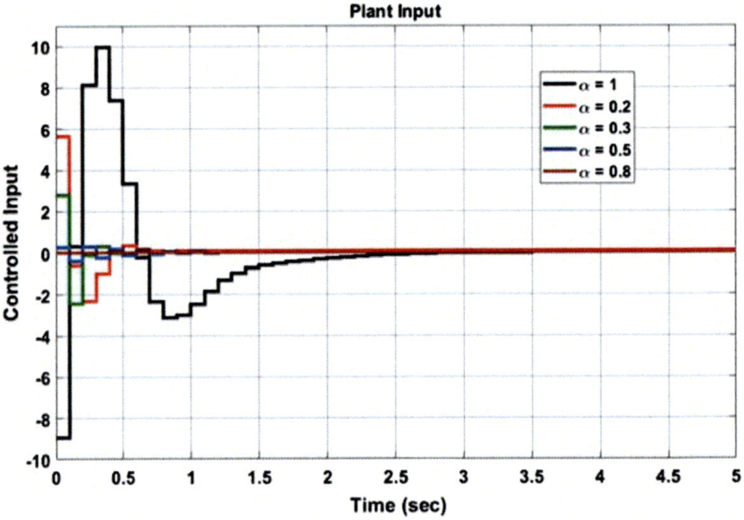

Fig. 4.5 Controlled input for $\alpha = 0.2, 0.3, 0.5, 0.8, 1$

Table 4.1 summarizes the essential attributes obtained from these plots. From Table 4.1, it can be observed that each FMPC response is better compared to the traditional MPC with $\alpha = 1$. Among these values of α, for $\alpha = 0.3$ the FMPC gives the best result for POAC system.

Let us check the robustness of this FMPC corresponding to the FO model with $\alpha = 0.3$. The capacity of a system to oppose change without adjusting initial stable

Table 4.1 Comparison table with different α values

Specifications	Settling time (Sec)		Oscillations	
α	$x(t)$	$\theta(t)$	$x(t)$	$\theta(t)$
0.2	0.49	0.84	No	Yes
0.3	**0.42**	**0.64**	**No**	**Yes**
0.5	1.03	1.40	Yes	Yes
0.8	1.01	1.77	Yes	Yes
1	3.15	3.55	No	Yes

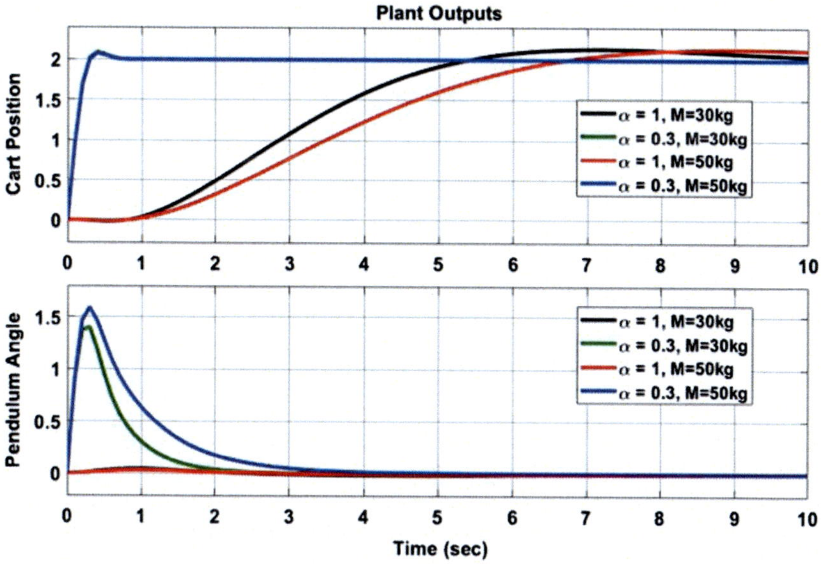

Fig. 4.6 Controlled output for $\alpha = 0.3$ and $M = 50$ Kg, $\alpha = 0.3$ and $M = 30$ Kg

condition is robustness [11]. The masses (Cart and Pendulum) may change, and so we consider two cases:

(i) changing cart mass, and
(ii) changing pendulum mass.

4.1.1 *Case 1: With Changing Cart Mass*

Consider the mass of the cart at one instant to be $M = 30$ kg and at another instant to be $M = 50$ kg, and let us design the FMPC for $\alpha = 0.3$. It is observed in Figs. 4.6 and 4.7, that the FMPC settles much faster as compared to traditional MPC whenever there is a change in the mass of the cart and hence FMPC is robust to change in the value of cart mass M.

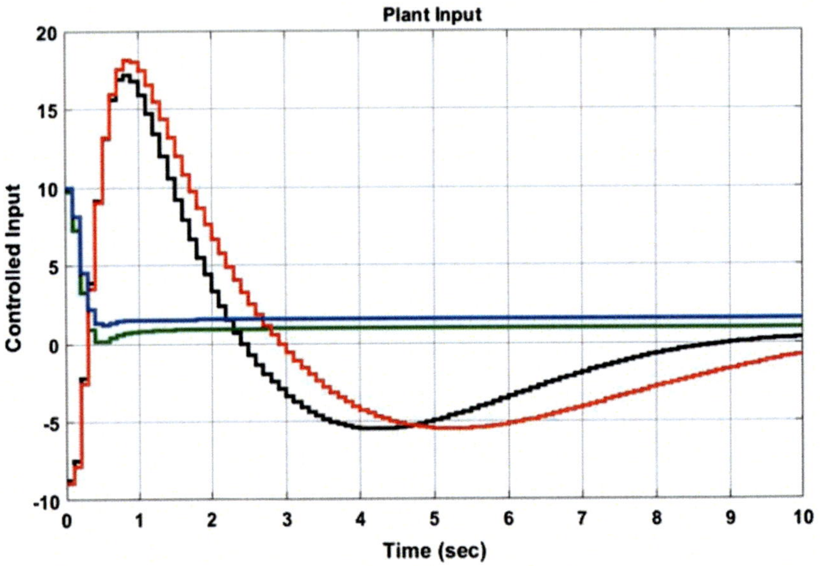

Fig. 4.7 Controlled input for $\alpha = 0.3$ and $M = 50\,\text{Kg}$, $\alpha = 0.3$ and $M = 30\,\text{Kg}$

4.1.2 *Case 2: With Changing Pendulum Mass*

Let us take the mass of the pendulum at one instant to be $m = 2\,\text{Kg}$ and at another time to be $m = 5\,\text{Kg}$ and design the FMPC for $\alpha = 0.3$. It is observed in Figs. 4.8 and 4.9 that the FMPC response settles fast as compared to traditional MPC to the changes in the pendulum mass, and hence it shows that FMPC is robust to change in the value of mass m. The FMPC for POAC system is designed and the stability is achieved, and it is also observed that FMPC control works pretty well for inverted pendulum on a cart (POAC) system. The proposed FMPC shows better results in all control aspects compared to the traditional MPC. Further, the FMPC is robust to the system parameter variations for POAC system and the conventional MPC to the changes of the mass of the pendulum.

4.2 Design of FMPC for 2-D Gantry Crane System

Most of the real-world systems are fractional, and IO models are localizations of FO models [12–15]. The classical IO Models are widely accepted to design any control system over FO model due to the lack of solution available [16]. However, the research work in the area of FC picked up in the past few decades. To get the physically significant form, a FO model is required to approximate the system using methods such as Oustaloup recursive approximation technique out of the various methods

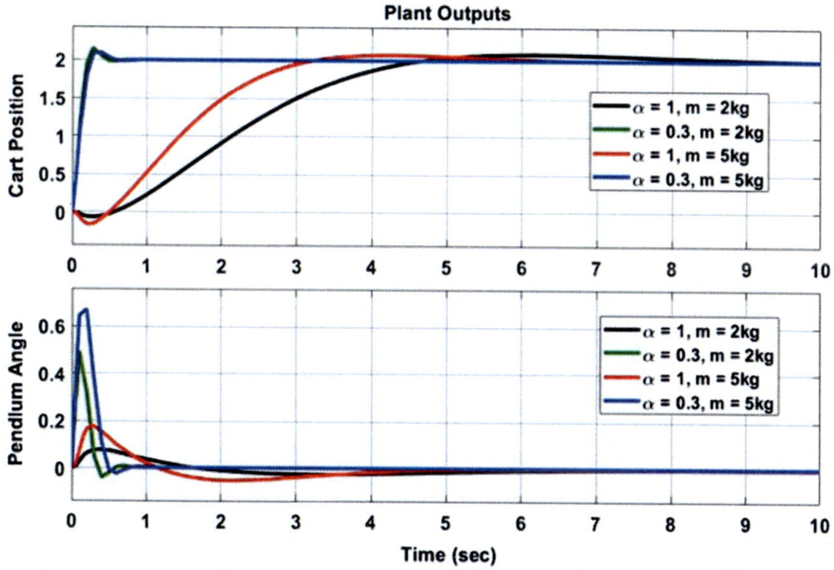

Fig. 4.8 Controlled output for $\alpha = 0.3$, $m = 2\,\text{Kg}$, and $\alpha = 0.3$, $m = 5\,\text{Kg}$

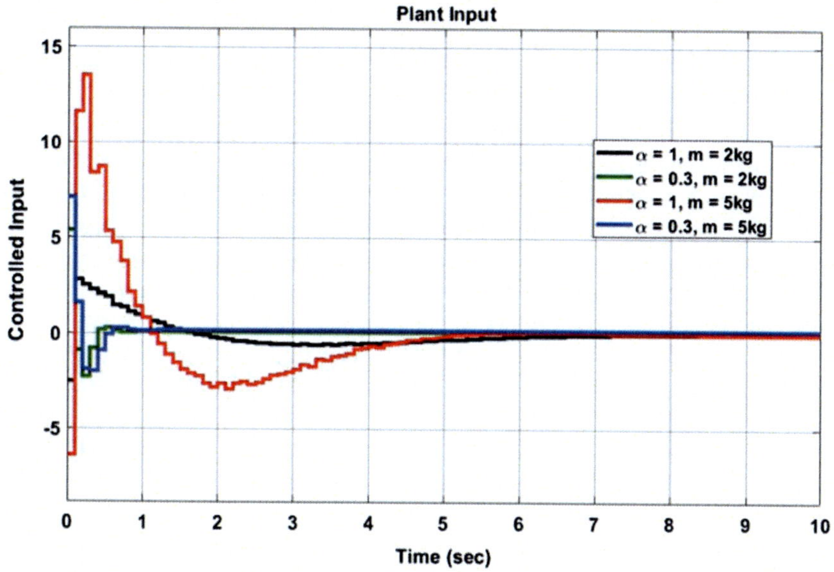

Fig. 4.9 Controlled output for $\alpha = 0.3$, $m = 2\,\text{Kg}$, and $\alpha = 0.3$, $m = 5\,\text{Kg}$

available for FO model approximation [17, 18]. The IO model of 2-D Gantry crane system is already converted to FO model in Chap. 3. For physical realization, this FO model is approximated using the Oustaloup recursive approximation technique

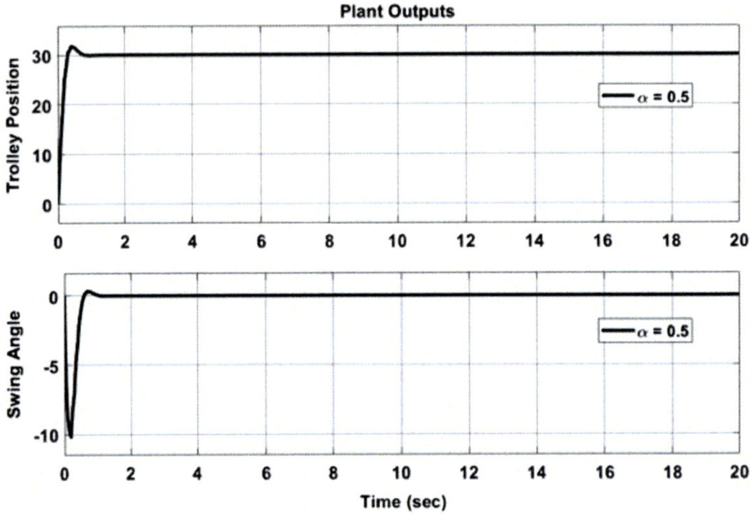

Fig. 4.10 Controlled output for $\alpha = 0.3$

and the MPC is designed. The implementation of the FMPC controller is done for the first time for 2-D Gantry crane system. The modeling is done in the Windows 8 environment using MATLAB MPC toolbox [10]. Figure 4.1 represents the structure of the MPC toolbox available in MATLAB.

The 2-D Gantry crane system is a Single Input Multiple Outputs (SIMO) system. The single input acts on the trolley and the two outputs viz. position and the swing angle are as shown in Fig. 4.1. The two system variables are the masses of load and trolley. The trolley position x(t) and swing angle θ need to be controlled in minimum time with no overshoot and no oscillations. The MPC toolbox is used to design the MPC for $\alpha = 0.3$, and with a sampling interval of 0.1 Sec, prediction horizon of 30 and control horizon of 6, and the output is shown in Figs. 4.10.

Clearly, $\alpha = 1$ gives an IO model for 2-D Gantry crane system. The MPC toolbox is used to design the MPC for the system for the same specifications as $\alpha = 0.3$. The output of the system for $\alpha = 0.3, 0.5, 0.8, 1$ is shown in Fig. 4.11.

It is obvious from Figs. 4.10 and 4.11 that the settling time of MPC designed for the FO Model of 2-D Gantry crane is greatly improved compared to the MPC designed for integer MPC.

For $0 < \alpha \leq 1$, these models correspond to the FO models of the 2D gantry system. The controller designed for these values of α improves the system output response. Hence, FMPC control design method for 2D gantry system can be opted to control the system response in the minimum amount of time with less overshoot. The comparisons are shown in Table 4.2 to show the superiority of the FMPC controller design method in terms of settling time and overshoot in the controlled responses. From the table, it is noticed that the FMPC controller performs better compared to the traditional MPC controller and FMPC improves the transient response. When

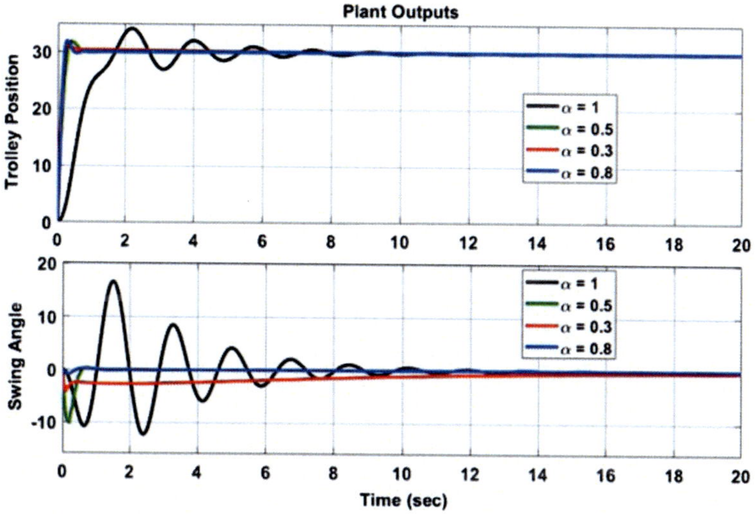

Fig. 4.11 Control outputs comparison of 2-D gantry crane system for $\alpha = 0.3, 0.5, 0.8, 1$

Table 4.2 Comparison table with different α values

Specification	Settling time (Sec)		Overshoot		Oscillations	
α	$x(t)$	$\theta(t)$	$x(t)$	$\theta(t)$	$x(t)$	$\theta(t)$
0.3	0.8	16	No	No	No	No
0.5	**0.8**	**3.5**	**No**	**No**	**No**	**No**
0.8	0.8	5	Yes	Yes	No	Yes
1	3.8	4	Yes	Yes	Yes	Yes

$\alpha = 0.5$, the corresponding model is a valid FO model of the considered system. The objective of designing FMPC for the considered robotic system, i.e., 2D gantry, is achieved. The FMPC control improves the transient response significantly. The proposed method of FMPC control gives improved response in all control aspects compared to traditional MPC.

Next, we study the FMPC control method's robustness of the FO model corresponding to $\alpha = 0.5$ for the 2-D Gantry crane system. Two scenarios are considered: (i) variation in the cart (trolley) mass, and (ii) variation in the load mass. Consider that the mass of cart is changed from $M = 10$ Kg to 50 Kg. Let us design the FMPC controller for the fractional model of $\alpha = 0.5$. Figure 4.12 shows the response wherein it can be observed that the controller corresponding to FMPC has perfect settling time and swing angle is not oscillating with zero overshoot in both position and swing angle when the mass of the trolley is changing.

If we observe the same case on integer model of 2D gantry crane system, the response corresponding to traditional MPC corresponding to $\alpha = 1$, the settling time is weak with oscillations and there is overshoot when the mass of the cart is changed.

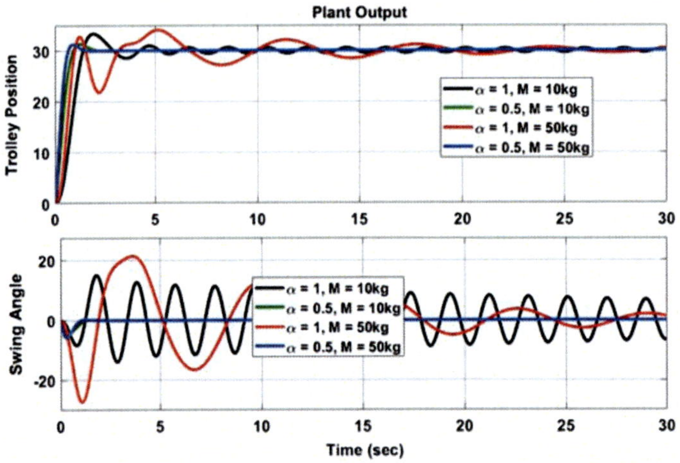

Fig. 4.12 The output response of the FMPC and MPC controllers for trolley mass M values

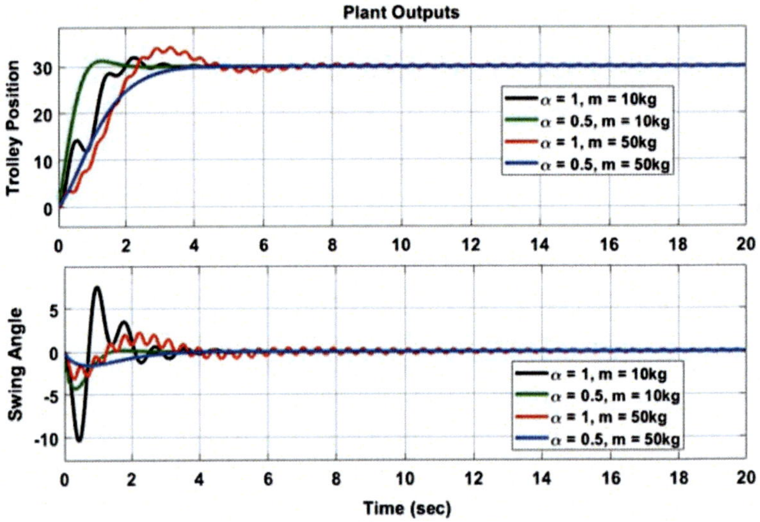

Fig. 4.13 The output response of the FMPC and MPC controllers for different load mass m values

For a higher value of trolley mass, traditional MPC will have higher oscillation and overshoot. Consider that the load mass is changed from $m = 10\,\text{Kg}$ to $50\,\text{Kg}$. Let us design the FMPC controller for the fractional model with $\alpha = 0.5$. Figure 4.13 represents the output response wherein it is observed that the controller based on FMPC has a good settling time. When compared to traditional MPC, there is almost no oscillations and zero overshoot for both position and swing angle with change in load mass. The summary of robustness is provided in Tables 4.3 and 4.4.

Table 4.3 Summary table with the trolley mass M changing

Specifications	Mass (Kg)	Settling time (Sec)		Overshoot		Oscillations	
α	M	$x(t)$	$\theta(t)$	$x(t)$	$\theta(t)$	$x(t)$	$\theta(t)$
0.5	10	0.66	14.4	No	No	No	No
0.5	50	0.75	9.2	No	No	No	No
1	10	8.9	19.8	Yes	Yes	Yes	Yes
1	50	9.5	90+	Yes	Yes	Yes	Yes

Table 4.4 Summary table with the load mass m changing

Specifications	Mass (Kg)	Settling time (Sec)		Overshoot		Oscillations	
α	M	$x(t)$	$\theta(t)$	$x(t)$	$\theta(t)$	$x(t)$	$\theta(t)$
0.5	10	0.66	17.4	No	No	Yes	No
0.5	50	4.3	7.18	No	No	No	No
1	10	5.65	7.75	Yes	Yes	Yes	Yes
1	50	28.5	70+	Yes	Yes	Yes	Yes

Table 4.5 Comparison with existing control methods for 2D gantry crane system

Paper	Settling time (sec)		Methodology
	Trolley	Swing angle	
Kolonic, F. et al. [22]	~3 sec	~4 sec	Tensor product
Tuan, L., A. et al. [23]	~5 sec	~6 sec	Sliding mode controller
Arpacia, H. et al. [24]	~6 sec, ~6 sec, ~8 sec	~6 sec, ~6 sec, ~6 sec	Fractional PID, ANFIS controller, PID controller
Wahyadi, J. J. et al. [25]	4.36 sec	~4 sec	NCTF control
Shebel, A. et al. [26]	~10 sec	–	Fuzzy-PD controller
Proposed controller	~0.8sec	~3.5 sec	**Fractional model predictive controller**

The FMPC design for 2-D Gantry crane system is robust and the performance is excellent in all aspects compared to the traditional MPC designs. Comparing the results from existing controllers for 2-D Gantry crane system (refer Table 4.5), it is inferred that the proposed FMPC controller performs better than the existing ones.

4.3 FMPC for Missile Launching Pad

A missile launch vehicle carries one or more ground-to-ground or ground-to-air missiles, along with the control (personnel) and equipment needed to organize and execute the launch of such missiles. Many of the launch vehicles are manually controlled. For proper launch of missiles, the angular movements of the launch vehicles need to be controlled. A control mechanism using the FMPC to control the angles of the missile launch vehicle is proposed. Some of the control methods of a missile launch vehicle are already there. In [19], an adaptive sliding mode controller based upon disturbance observer is proposed for launch vehicle, and a sliding mode control strategy for a launch vehicle is developed in [20]. The dynamics of the missile launch vehicle is discussed [21].

The mathematical modeling of launching vehicle is obtained from Euler-Lagrange formulation (Chap. 2) and expressed in its fractional equivalent model (Chap. 3). The model is approximated using Oustaloup-Recursive-Approximation for different FO values of α. These FO models are then considered for MPC. Using simulations, it is observed that the FMPC for the launching vehicle gives a better response as compared to a regular MPC. As per the knowledge of the authors, FMPC controller is designed for the first time for missile launching vehicle (MLV) system. To design the controller, a MATLAB toolbox of MPC [10] is opted for as shown in Fig. 4.14. Let us design the MPC using the above MPC toolbox with a sampling interval of 0.2 sec, prediction horizon 30 and control horizon 6.

The aim is to control the angle one (θ_1) at 30 degrees and angle two (θ_2) at 45 degrees. The response corresponding to these specifications is shown in Fig. 4.15.

From Fig. 4.15, it can be noticed that the FMPC gives the desired output in minimum time and can be designed to control the angles of the missile launch vehicle. Let us consider the integer or regular model of the MLV system. Considering the specifications of the MPC controller to be same as that of $\alpha = 0.1, 0.2, 0.5, 0.8$, Figs. 4.15 and 4.16 give the output response of the non-integer and regular model of the MLV system. It can be observed from Fig. 4.15 that there are no oscillations in the FMPC of MLV system for $\alpha = 0.1$ whereas the integer or regular model predictive controller gives oscillations and also overshoot. Let us compare in detail the performances for the MPC for $\alpha = 0.1, 0.2, 0.5, 0.8$ and 1. Figure 4.17 show the detailed controlled comparison analysis for MLV system. The FMPC does not have any oscillations in

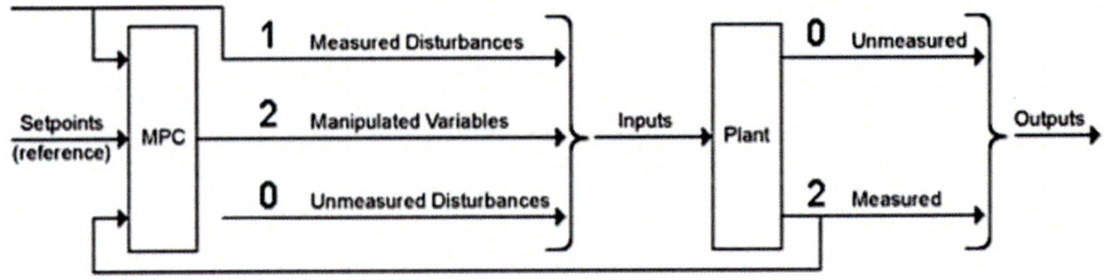

Fig. 4.14 MPC toolbox structure for MLV system

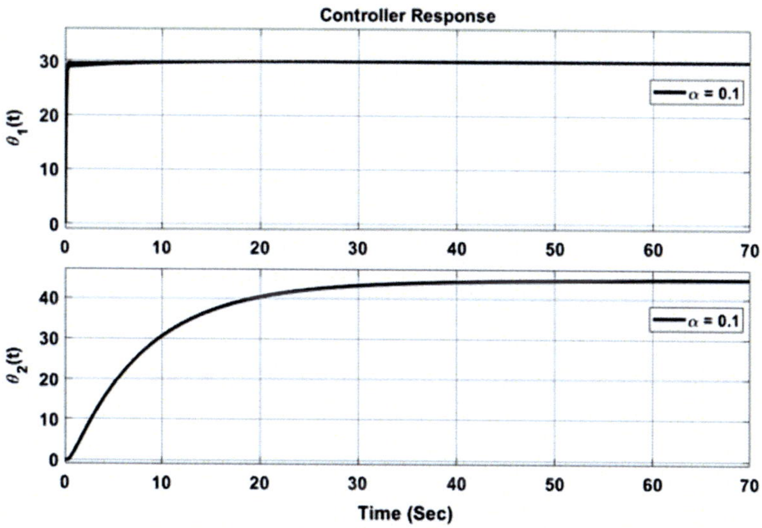

Fig. 4.15 The control output of the fractional MLV system for $\alpha = 0.1$

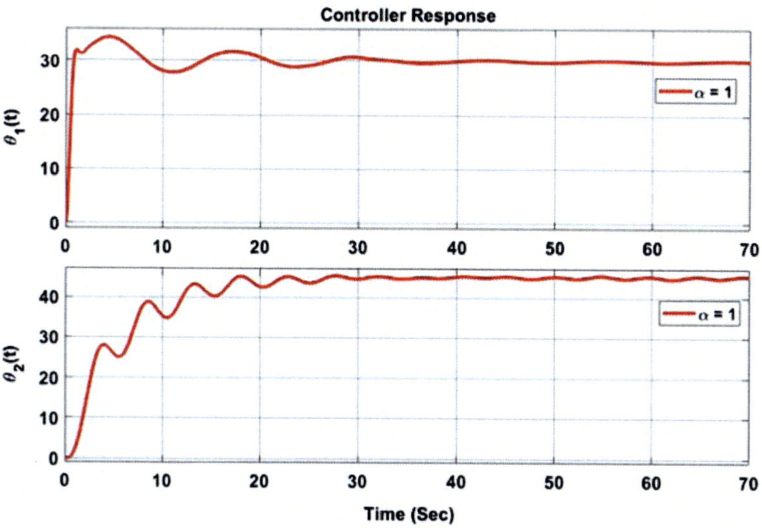

Fig. 4.16 The control output of the integer MLV system for $\alpha = 1$

its output and gives the desired output in minimal time as compare to the traditional MPC (i.e., $\alpha = 1$). It can be summarised from the comparison of various outputs of the controllers that the FMPC performs better than MPC. The comparison summary Table 4.6 is also provided for a better understanding of the FMPC versus MPC.

The main goal of designing a FMPC for missile launching vehicle is achieved, and it is seen that FMPC works well in all control aspects. Table 4.6 presents the

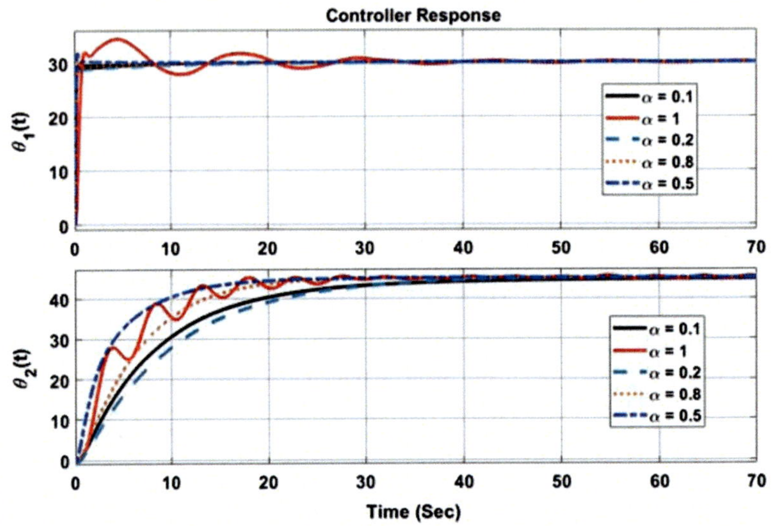

Fig. 4.17 Comparison of controlled outputs of MLV system for $\alpha = 0.1, 0.2, 0.5, 0.8, 1$

Table 4.6 Comparison table with different α values

Specification	Settling time (Sec)		Overshoot		Oscillations	
α	$\theta_1(t)$	$\theta_2(t)$	$\theta_1(t)$	$\theta_2(t)$	$\theta_1(t)$	$\theta_2(t)$
0.1	12	47	No	No	No	No
0.2	30	35	No	No	No	No
0.5	**6**	**23**	**Low**	**No**	**No**	**No**
0.8	25	30	No	No	No	No
1	50	45	Yes	Yes	Yes	Yes

detailed advantages of using FMPC. From the table, it is concluded that FMPC is a better controller than traditional MPC and with $\alpha = 0.5$, FMPC controller gives the best result compared to conventional MPC controller for missile launching vehicle system. Section 4.1 presents the MPC design for the fractional model of POAC system and finds that with $\alpha = 0.3$, the FO model gives the best response. The robustness of the considered fractional model of POAC is also checked for the variation of system parameters. Section 4.2 presents the MPC design for a fractional model of 2D gantry crane system and finds that with $\alpha = 0.5$, the FO model gives the best response and suggests that the model corresponding to $\alpha = 0.5$ be the equivalent fractional model of this system. A comparison of the controller response with existing control methods is shown in Table 4.5. The robustness of the considered fractional model of 2D gantry crane system is also checked for variation of system parameters. Section 4.3 presents the MPC design for the fractional model of MLV system and finds that with $\alpha = 0.5$, the FO model gives the best response and suggests that the model corresponding to

$\alpha = 0.5$ is the equivalent fractional model of this system. In the next chapter, a pole placement method of the controller design is proposed for the FO model of a single link robotic manipulator.

4.4 Adaptive Fractional Model Predictive Control (AFMPC)

In the systems thus far, we studied linear dynamics and assumed that the considered systems have fixed values of system parameters, so the plant dynamics doesn't change and the state matrix 'A' is constant. Adaptive controllers are able to adapt to system uncertainties and give the best possible controlled response [27, 28]. Therefore, it is appropriate at this stage to consider systems whose dynamics changes with time due to varying nature of certain system parameters [29–32].

To control different types of systems we have used MPC and FMPC controllers in this chapter so far, but now we will let some system parameters to vary at certain arbitrary time instants. That is, the state matrix 'A' is changing too. The traditional MPC and FMPC controller are not effective at handling the varying dynamics because of using constant internal plant fractional model, and hence the natural requirement is to explore Adaptive FMPC (AFMPC). The AFMPC provides a new linear plant model at each time step as the operating condition changes and therefore it is able to make more accurate predictions for the new operating conditions. Therefore, to deal with the change in the plant dynamics [33], we are going to use AFMPC and the block diagram of operation is shown in Fig. 4.18.

With a system model in place, we run a set of forecasts of this model forward in time for different actuation strategies, and optimize the control input u over a short interval of time and this essentially determines the immediate next control action based on that optimization. Then we reinitialize our optimization to find our next control inputs u and such a process essentially keeps on happening to find the best

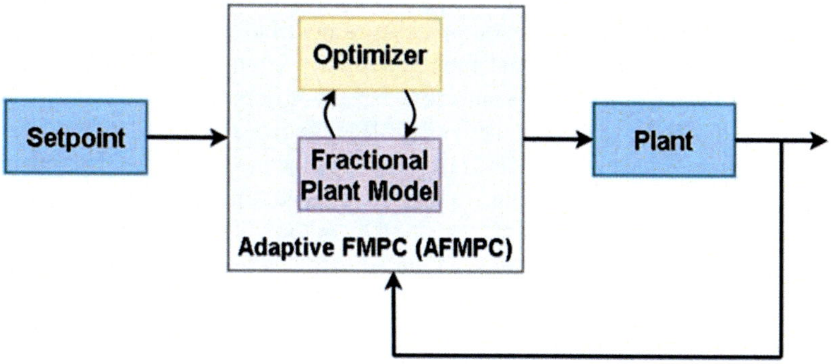

Fig. 4.18 AFMPC block diagram

optimized control input. The basic idea is, let us say, we have some time t and to obtain a desired output y_d, a control signal u is available.

Let us say that there is some kind of a set point, a goal, that we are trying to reach. We want our system to go to the desired location and let us say that it is started at some initial state x and at time k, and let us say that $y = x$, just to keep it simple. Then we start the system and initialize it at time k with state $x(k)$ and run the optimization procedure over the input u to try to find what is the best u over a certain short interval of time that may fulfill the desired objective. So, there is a window of time and lets call it as the horizon in MPC over which are optimize the input u. Then we lock in the first value of the optimized control law and then check where the system response goes. If the system is still not closer to the reference then we shift our horizon over some small time and then we re-optimize to get a different optimal value of control law and lock that in and see where the system response goes. So over time, we can essentially watch our system step forward and at every new time step we reassess the optimizing control input u to obtain desired control. For a linear system, we determine the optimal control signal using LQR control and use it for MPC short time optimization and those will compensate over time. In case of a FMPC, the same inputs are applied as in the case of the traditional MPC. The only difference is that the FMPC takes a fractional equivalent plant model for further optimization and further the same steps would be followed to get a good response from the FMPC. The optimization block shown in Fig. 4.19 is the heart of MPC.

The optimizer block is supposed to minimize a cost function

$$J = \sum_{i=1}^{p} w_e e_{k+i}^2 + \sum_{i=0}^{p-1} w_{\Delta u} \Delta u_{k+i}^2, \tag{4.1}$$

where w is the weighted sum, e is the predicted output and u is the control input, meant to minimize the deviation between the desired and predicted outputs.

The cost function of this optimization is represented as weighted square sum of predictive errors and the actual input u. At the current time step, the MPC solves the optimization problem over the prediction horizon. The predicted path over the smallest J gives the optimal solution, and therefore determines how much input is required to make the prediction as close as possible to the reference or desired position. The adaptive fractional model predictive controller (AFMPC) block has the same inputs as the FMPC controllers but the difference is that it will take one extra input that is the varying plant dynamic fractional order equivalent model at each instant of time. An adaptive FMPC is required to take the same objective function on which the FMPC works and then we need to proceed further to make the prediction as close as possible to the reference position by optimizing the cost function of (4.1).

Next, we study the effectiveness of the AFMPC on three systems: (i) POAC, (ii) 2 D Gantry Crane, and (iii) MLV systems.

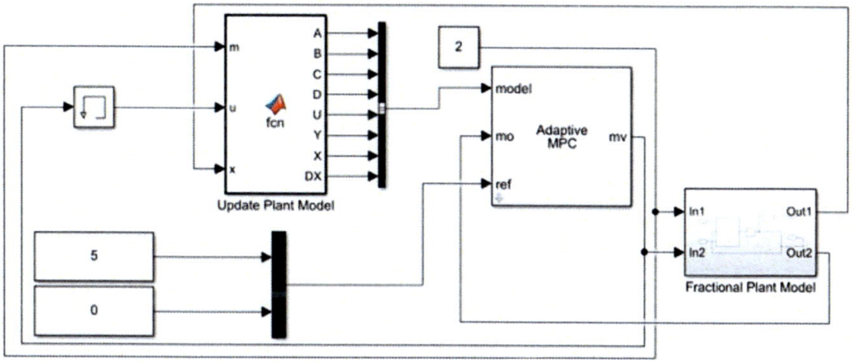

Fig. 4.19 AFMPC simulink model

4.4.1 For POAC System

The design of AFMPC for POAC system provided using MATLAB/Simulink is shown in Fig. 4.19. The steps to develop such a model can be described in simple terms as follows. Open a new Simulink model and define a FO plant corresponding to POAC system for $\alpha = 0.3$ as this is the best FO model of POAC system) with input as a constant force and output as cart position and pendulum angle. Let us assume that the dynamics change with pendulum mass 'm' which is applied as a input to the plant. Let the mass of the pendulum is initially set at 2 kg. Next, we connect the adaptive model predictive controller block which is available in MPC toolbox library and has the same inputs as of the traditional MPC block, but the only difference is that it takes one more input, that is, the changing plant model at each time. To implement with adaptive FMPC, we simply start with FMPC controller designed in the previous sections for a constant pendulum mass '$m = 0.5$ kg'.

We already have the MPC controller object and hence we can see the design parameters such as the prediction and control horizons. The adaptive FMPC block requires discrete time plant model and hence it is required to convert the continuous time plant model which is used by FMPC into a discrete time model. Next we need to put this FMPC object to AFMPC and provide the controller with a fractional equivalent plant model which is updated at every time for the present operating condition. The updated plant model block (with inbuilt function shown in Appendix A.2) takes care of the calculations. Next, connect the inputs and outputs for this block and we can change different values of pendulum mass 'm' and see how this controller handles the varying plant dynamics. In the previous section, the FMPC controller designed for an operating condition of '$m = 0.5$' worked well, but for different pendulum masses the response of the controller is not as expected. With AFMPC we get good controller performance even when pendulum mass is changed to some other values as shown in Fig. 4.20. We can safely conclude that the AFMPC can deal with the changing plant dynamics and can successfully control the POAC

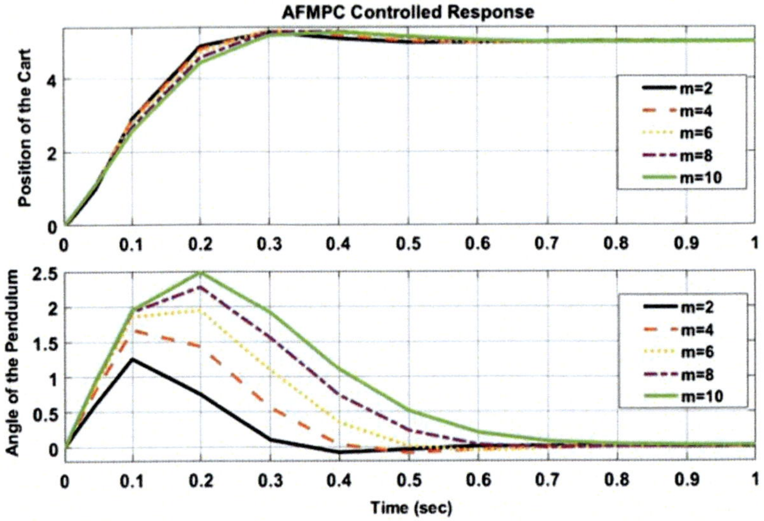

Fig. 4.20 Adaptive fractional model predictive controller (AFMPC) response for changing POAC plant dynamics

system. So, we have designed an AFMPC controller and can run several simulations to evaluate the performance.

4.4.2 For 2D Gantry Crane System

Let us again consider a 2D Gantry Crane System whose dynamics change with time due to varying mass 'm' of the load. In Sect. 4.2, we have learned linear system dynamics and assumed that the load has a constant value of mass 'm'. Therefore, the plant dynamics do not change and the state matrix 'A' is constant. To control such types of systems, we have use MPC and FMPC controllers, but now we let the mass of the load vary as the cart moves. So, the state matrix 'A' changes too. A traditional MPC and FMPC controller is not effective in handling the varying dynamics as it uses a constant internal plant fractional model and hence Adaptive FMPC (AFMPC) is an appropriate choice. The AFMPC lets us provide a new linear plant model at each time step as the operating condition changes and therefore it makes more accurate predictions for the new operating conditions. So to deal with the change in the plant dynamics we use AFMPC. We follow the same steps as discussed in the case of the POAC system for designing the AFMPC controller for the 2D gantry crane system.

We conclude from Fig. 4.21 that the AFMPC can deal with the changing plant dynamics and can successfully control the 2D Gantry Crane system. So, we have designed an AFMPC controller for 2D Gantry crane system and can run several simulations to evaluate the performance.

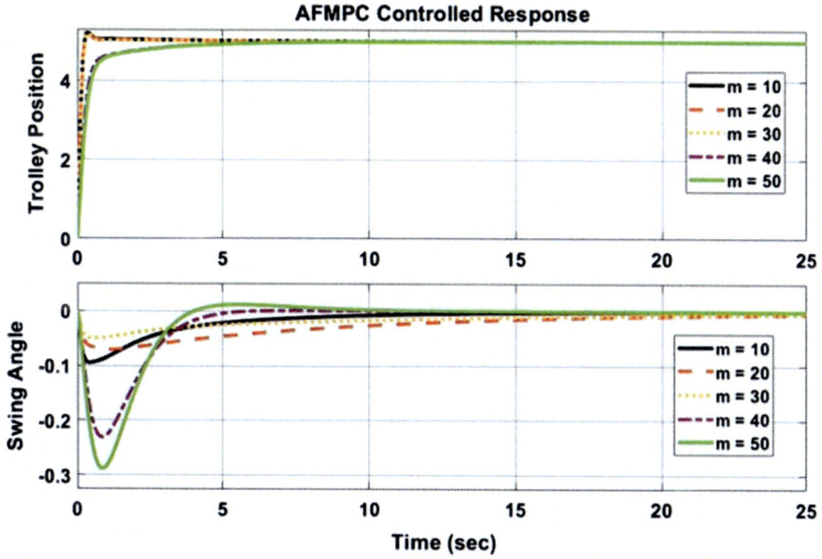

Fig. 4.21 Adaptive fractional model predictive controller (AFMPC) response for changing 2D gantry crane plant dynamics

References

1. Deng, Z., Cao, H., Li, X., Jiang, J., Yang, J., Qin, Y.: Generalized predictive control for fractional order dynamic model of solid oxide fuel cell output power. J. Power Sources **195**(24), 8097–8103 (2010)
2. Domek, S.: Switched state model predictive control of fractional-order nonlinear discrete-time systems. Asian J. Control **15**(3), 658–668 (2013)
3. Rhouma, A., Bouani, F., Bouzouita, B., Ksouri, M.: Model predictive control of fractional order systems. J. Comput. Nonlinear Dyn. **9**(3), 031011 (2014)
4. Sopasakis, P., Sarimveis, H.: Stabilising model predictive control for discrete-time fractional-order systems. Automatica **75**, 24–31 (2017)
5. Zou, Q., Jin, Q., Zhang, R.: Design of fractional order predictive functional control for fractional industrial processes. Chemom. Intell. Lab. Syst. **152**, 34–41 (2016)
6. Zhang, R., Zou, Q., Cao, Z., Gao, F.: Design of fractional order modeling based extended non-minimal state space MPC for temperature in an industrial electric heating furnace. J. Process Control **56**, 13–22 (2017)
7. Singh, A.P., Agarwal, H., Srivastava, P., Naidu, Praveen: A robust fractional model predictive control design. Progress Fract. Differ. Appl. **5**(3), 217–223 (2019)
8. Singh, A., Agrawal, H.: A fractional model predictive control design for 2-d gantry crane system. J. Eng. Sci. Technol. **13**(7), 2224–2235 (2018)
9. Singh, A.P., Agrawal, H., Srivastava, P.: Robust fractional model predictive controller (FMPC) design for under-actuated robotic systems. Int. J. Control Autom. **11**(7), 2224–2235 (2018)
10. Bemporad, A., Morari, M., Ricker, N.L.: Model Predictive Control Toolbox User's Guide. The mathworks (2010)
11. Alippi, C.: Intelligence for Embedded Systems. Springer, Switzerland (2014)
12. Sabatier, J.A.T.M.J., Agrawal, O.P., Machado, J.T.: Advances in Fractional Calculus 4(9). Springer, Dordrecht (2007)

13. Tepljakov, A., Petlenkov, E., Belikov, J., Halas, M.: Design and implementation of fractional-order PID controllers for a fluid tank system. In: 2013 American Control Conference. IEEE, pp. 1777–1782 (2013)
14. Singh, A.P., Deb, D., Agarwal, H.: On selection of improved fractional model and control of different systems with experimental validation. Commun. Nonlinear Sci. Numer. Simul. **79**, 104902 (2019)
15. Kaslik, E., Sivasundaram, S.: Nonlinear dynamics and chaos in fractional-order neural networks. Neural Netw. **32**, 245–256 (2012)
16. Chen, Y., Petras, I., Xue, D.: Fractional order control-a tutorial. In: American Control Conference. IEEE, pp. 1397–1411 (2009)
17. Vinagre, B.M., Podlubny, I., Hernandez, A., Feliu, V.: Some approximations of fractional order operators used in control theory and applications. Fract. Calc. Appl. Anal. **3**(3), 231–248 (2000)
18. Tavazoei, M.S., Haeri, M.: Unreliability of frequency-domain approximation in recognising chaos in fractional-order systems. IET Signal Process. **1**(4), 171–181 (2007)
19. Wang, Z., Wu, Z., Du, Y.: Adaptive sliding mode backstepping control for entry reusable launch vehicles based on nonlinear disturbance observer. Proc. Inst. Mech. Eng. Part G: J. Aerosp. Eng. **230**(1), 19–29 (2016)
20. Ni, S., Shan, J.: Smooth second-order nonsingular terminal sliding mode control for reusable launch vehicles. Int. J. Intell. Comput. Cybern. **7**(1), 95–110 (2014)
21. Singh, A.P., Agrawal, H.: On modeling and analysis of launch vehicle system. In: Proceedings of the International Conference on Signal, Networks, Computing, and Systems. Springer, New Delhi, pp. 273–279 (2016)
22. Koloni, F., Poljugan, A., Petroviĉ, I.: Tensor product model transformation-based controller design for gantry crane control system-an application approach. Acta Polytechnica Hungarica **3**(4), 95–112 (2006)
23. Tuan, L.A., Kim, G.H., Lee, S.G.: Design of sliding mode controller for the 2D motion of an overhead crane with varying cable length. In: Journal of Mechanism and Machine Theory (2016)
24. Arpaca, H., Özgüvenb, Ö.F.: ANFIS & PID controller design and comparison for overhead cranes. Indian J. Eng. Mater. Sci. **18**, 191–203 (2011)
25. Wahyudi, Jalani, J., Muhida, R., Salami, M.J.E.: Control strategy for automatic gantry crane systems: a practical and intelligent approach. Int. J. Adv. Robot. Syst. **4**(4), 46 (2007)
26. Shebel, A., Maazouz, S., Mohammed Abu, Z., Mohammad, A., Ayman Al, R.: Design of fuzzy PD-controlled overhead crane system with anti-swing compensation. Engineering (2011)
27. Patel, R., Deb, D., Dey, R., E. Balas, V.: Introduction to Adaptive Control. Intelligent Systems Reference Library, pp. 53–65 (2019). https://doi.org/10.1007/978-3-030-18068-3_5
28. Malepati, B., Vijay, D., Deb, D., Manjunath, K.: Parameter validation to ascertain voltage of Li-ion battery through adaptive control. Adv. Intell. Syst. Comput. **21–31** (2018). https://doi.org/10.1007/978-981-13-1966-2_3
29. Patel, R., Deb, D., Dey, R., Balas, E.V., : Adaptive control of single chamber two-population MFC. IN: Intelligent Systems Reference Library **81–89** (2019). https://doi.org/10.1007/978-3-030-18068-3_7
30. Patel, R., Deb, D., Dey, R., Balas, E.V., : Adaptive and intelligent control of microbial fuel cells. In: Intelligent Systems Reference Library (2020). https://doi.org/10.1007/978-3-030-18068-3
31. Nath, A., Deb, D., Dey, R.: An augmented subcutaneous type 1 diabetic patient modelling and design of adaptive glucose control. J. Process Control **86**, 94–105 (2020). https://doi.org/10.1016/j.jprocont.2019.08.010
32. Nath, A., Deb, D., Dey, R.: Robust observer based adaptive control of blood glucose in diabetic patients. Int. J. Control **1–0** (2020). https://doi.org/10.1080/00207179.2020.1750705
33. Deb, D., Tao, G., Burkholder, J.: An adaptive inverse compensation scheme for signal-dependent actuator nonlinearities. In: 2007 46th IEEE Conference On Decision And Control (2007). https://doi.org/10.1109/cdc.2007.4434743

Chapter 5
Modeling, Stability and Fractional Control of Single Flexible Link Robotic Manipulator

In the previous chapter, an FMPC controller is explored to control the FO model of robotic manipulators. This chapter deals with the fractional order modeling, stability analysis and control of a single flexible link robotic manipulator (SFLRM). The control law is derived using pole placement (PP) method. This paper uses Mittag-Leffler function for the analysis of SFLRM in the time domain. The stability analysis of the fractional model is carried in a transformed Ω-Domain and from the analysis, it is observed that the response of the fractional model of SFLRM robotic manipulator is stable. The main motive behind this analysis is to understand the fractional behavior of SFLRM and it is well known from the literature that most of the real-world systems have their own fractional behavior. Furthermore, a real-time SFLRM setup is considered to validate the results obtained and it is found that the control law suggested by PP method improves the settling-time of SFLRM.

Modeling of any system plays a major role in understanding the behavior of the system [1]. In a large number of systems, the more accurate the mathematical representation of the actual dynamical system is [2, 3], the more precise control of the same is possible. However, the accuracy of the mathematical model is always an important factor in many different applications. Modeling of systems utilizing the concept of fractional calculus gives a more accurate system model. Fractional calculus has tremendous applications in nearly each and every field of science and technology as proven through literature over the years. Therefore, with a view to demonstrate the efficacy of fractional calculus theory, through a simple experiment, there is a need to focus on the areas of modeling, stability analysis and control of a single link robotic manipulator.

Fractional models have been used in diverse areas such as modeling of the electrochemical impedance for an accurate estimation of system parameters [4], for improving ethanol concentration [5], and for modeling of a diode [6]. Various other papers in the literature also demonstrate the superiority of the fractional model and control of a system; an algorithm is proposed to select an appropriate fractional-

A. P. Singh et al., *Fractional Modeling and Controller Design of Robotic Manipulators*, Intelligent Systems Reference Library 194,
https://doi.org/10.1007/978-3-030-58247-0_5

order model of a system to design controller [7] and also the merit of the proposed algorithm is tested on robotic manipulators to show the superiority of fractional calculus. Elsewhere, a fractional-order controller is designed to improve the transient response of pendulum-cart system [8] and gantry crane system [9]. Fractional modeling of SFLRM has been presented [10] using the fractional Laplace transformations for system modeling with an extensive history [11, 12]. The model obtained in the present work is in agreement with the states claimed elsewhere [13]. Once the fractional model is obtained, it has to go through a stability analysis for deeper insights. Various concepts of stability analysis of fractional-order (FO) systems are available [14–16]. This paper opts for the fractional stability analysis in Ω-domain [17].

Various control formulations to facilitate control of different systems are available in the literature. A few of these which are related to fractional controller design are fractional PID controller for twin rotor aerodynamics system [18], fractional control of multi-body systems [19] and general systems [20], pH neutralization process [21], pressure plant [22], magnetic levitation [23, 24] and redundant robots [25], and fractional sliding mode control of wind farm systems [26], and optimal control of linear fractional systems [27, 28]. Such controllers have proved to be efficient when compared to the integer counterparts. A fractional-order linear multi-agent system [29] using fractional calculus approach is found to provide improved system performance. A fractional PD^{μ} controller strategy [30] used to control LTI system with delay, gives an improved performance as compared with the traditional controller. The stability of a fractional-order sliding mode controller designed [31] for a class chaotic systems proves to be robust and reduces chattering phenomena. Recently, the fractional-order control method has been explored in the area of the traction system of the Electronic Vehicles [32]. An interesting survey paper [33] on fractional-order control strategy on different systems presents different examples where the fractional control strategy improves system performance. These outcomes prove the superiority of the fractional controller in improving the overall system performance significantly.

Various strategies are adopted in literature, to design control law for SFLRM. For a 2DOF flexible manipulator, a strategy to design LQR based robust control law is reported [34]. There are various papers available on sliding mode control of SFLRM; for effective closed-loop tracking as well as rigid dynamics of SFLRM [35], for better positional accuracy and robustness using second-order super twisting algorithm [36], and for better estimation of the tip deflection using discrete sliding Fourier transform [37]. Fractional sliding mode controller provides tremendous potential to improve the SFLRM system responses, a state estimation approach based fractional sliding mode observer is proposed [38]. A controller design method based on PID for SFLRM is presented [39] for input tracking and vibration separation, and for minimum vibration for optimizing the PID controller [40]. Neural network-based control is proposed to improve the performance of SFLRM; an active fuzzy logic-based vibration control is proposed to control the position of the tooltip with minimum vibrations [41], an adaptive neural network-based controller design is studied to improve the elastic deflection accuracy of the flexible manipulator [42], and an assumed mode method is proposed to analyze the flexible link by opting neural network control to track the trajectory of the link and minimize the vibrations [43]. Various other controller design solutions are available to control the flexible manipulator; an observer-based control using feedback linearization for trajectory tracking [44], and a hybrid passivity-based position control of the tip of a flexible manipulator [45]. A pole placement method of control analysis for a fractional model of SFLRM is not available in literature as per the authors' knowledge and therefore, this paper proposes a PP method of controller design for the fractional model of SFLRM to improve the system performance. The contributions of this work are

(i) Fractional model of SFLRM system which closely follows the properties of the actual model dynamics.
(ii) Region of convergence of SFLRM system.
(iii) Design of control law using the PP method for a fractional model of SFLRM which ensures a reduction in the settling time of the angular position of the link.
(vi) Experimental validation of the simulation results is ensured.

In this chapter, fractional modeling, fractional stability analysis and a control law using pole placement method are presented for the fractional model of SFLRM system. Furthermore, the analysis is carried on a real-time experimental setup of SFLRM and it is found that the control law designed for the fractional model improves the system performance. Overview of the paper is as follows. Section 5.2 discusses fractional-order modeling of the SFLRM system and its model validation. In Sect. 5.3, stability analysis of SFLRM system is carried out in Ω-plane, followed by the control law of the considered system, and experimental validation is provided in Sect. 5.4, followed by chapter summary.

5.1 Fractional Order Model of SFLRM and Its Validation

Let us consider a single link robotic manipulator as shown in Fig. 5.1.

The considered manipulator model is taken from the manual provided by Quanser Inc. [47]. This manipulator has one flexible link whose tool tip can rotate to and fro at an angle of 90°. The link length is $L = 0.42$ m with mass $m = 0.07$ kg, damping $B =$ 0.004 N.M/(rad/sec) and inertia $J = 2.08 \times 10^{-3}$ kg.m^2. The modeling equations

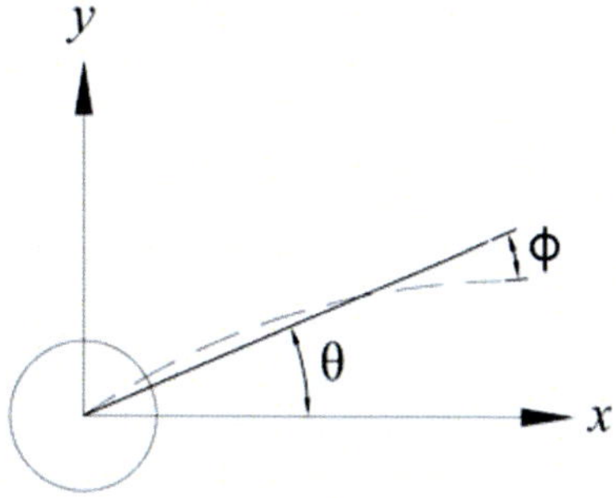

Fig. 5.1 Rotary flexible link schematic

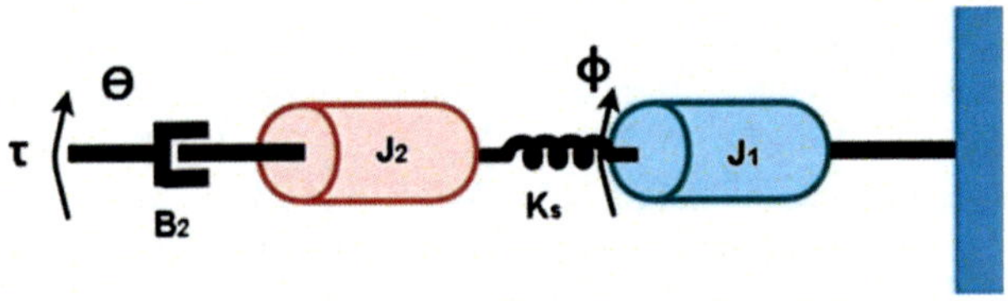

Fig. 5.2 Single flexible link mechanical model

are derived using Euler-Lagrange formulation and then linearized to get a linear state-space representation as

$$\dot{x} = Ax + Bu$$
$$y = Cx + Du, \tag{5.1}$$

where

$$A = \begin{pmatrix} 0 & 0 & 1 & 0 \\ 0 & 0 & 0 & 1 \\ 0 & 625 & -7.2 & 0 \\ 0 & -967.1 & 7.2 & 0 \end{pmatrix}, \quad B = \begin{pmatrix} 0 \\ 0 \\ -480.8 \\ 480.8 \end{pmatrix},$$

$$C = \begin{pmatrix} 1\,0\,0\,0 \\ 0\,1\,0\,0 \end{pmatrix}, \quad D = \begin{pmatrix} 0 \\ 0 \end{pmatrix}.$$

A mechanical equivalent model is shown in Fig. 5.2. The linear state space model is obtained by finding the Lagrangian equation of the system and then by using the Euler-Lagrange formulation [47]. The Lagrangian's equation can be found by calculating the difference between the kinetic energy and the potential energy of the system. The steps involved in deriving (1) is shown below. The total kinetic energy (KE) of the system using Lagrangian mechanics can be found to be

$$KE = \frac{1}{2}J_2\dot{\theta}^2 + \frac{1}{2}J_2(\dot{\theta} + \dot{\phi})^2,$$

and the potential energy (PE) is expressed as

$$PE = \frac{1}{2}K_s\phi^2,$$

where J_1 is the moment of inertia of the link and K_s is the measured stiffness. The Lagrangian L can be found to be

$$L = KE - PE = \frac{1}{2}J_2\dot{\theta}^2 + \frac{1}{2}J_1(\dot{\theta} + \dot{\phi})^2 - \frac{1}{2}K_s\phi^2.$$

Using this L in the Euler-Lagrangian formulation, we get

$$\frac{d}{dt}\frac{\partial L}{\partial \dot{\theta}} - \frac{\partial L}{\partial \theta} = \tau - B_2\dot{\theta}, \quad \frac{d}{dt}\frac{\partial L}{\partial \dot{\phi}} - \frac{\partial L}{\partial \phi} = 0.$$

Substituting the value of L, the Lagrangian, in the above two equations, we get the governing equations of motions as

$$(J_2 + J_1)\ddot{\theta} + J_1\ddot{\phi} + B_2\dot{\theta} = \tau, \quad J_1\ddot{\theta} + J_1\ddot{\phi} + K_s\phi = 0.$$

Now, separating the value of $\ddot{\theta}$ and $\ddot{\phi}$, we get

$$\ddot{\theta} = -\frac{B_2}{J_2}\dot{\theta} + \frac{K_s}{J_2}\phi + \frac{1}{J_2}\tau,$$
$$\ddot{\phi} = \frac{B_2}{J_2}\dot{\theta} + \frac{K_s(J_1 + J_2)}{J_1 J_2}\phi - \frac{1}{J_2}\tau.$$

Taking the above equations of $\ddot{\theta}$ and $\ddot{\phi}$, and casting into the linear state space model, the transfer function of this system is represented as

$$\begin{bmatrix}\theta(s)\\ \phi(s)\end{bmatrix} = \begin{bmatrix}\frac{480s^2+164470}{s^4+7.2s^3+967.1s^2+2467s}\\ \frac{-480s^2}{s^4+7.2s^3+967.1s^2+2467s}\end{bmatrix}, \tag{5.2}$$

where $\theta(s)$ is servo angle/Motor shaft position (deg.) and $\phi(s)$ is tip-deflection angle. The transfer function is converted to its equivalent FO using the definition of FO Laplace Transformation [11] assuming zero initial conditions:

$$\begin{bmatrix}\theta(s)\\ \phi(s)\end{bmatrix} = K\begin{bmatrix}\frac{480s^{2\alpha}+164470}{s^{4\alpha}+7.2s^{3\alpha}+967.1s^{2\alpha}+2467s^{\alpha}}\\ \frac{-480s^{2\alpha}}{s^{4\alpha}+7.2s^{3\alpha}+967.1s^{2\alpha}+2467s^{\alpha}}\end{bmatrix}, \tag{5.3}$$

where α represents a fractional value in between $0 < \alpha \leq 1$ and K is the gain required by the fractional order model to act in the same manner as the integer model responds. Clearly, if α, $K = 1$ then (5.3) represents (5.2). Let us take $\alpha = 0.5$ so as to simplify the mathematical analysis and for simplified numerical simulations, then the equivalent transfer function is

$$\begin{bmatrix}\theta(s)\\ \phi(s)\end{bmatrix} = K\begin{bmatrix}\frac{480s+164470}{s^2+7.2s^{1.5}+967.1s+2467s^{0.5}}\\ \frac{-480s}{s^2+7.2s^{1.5}+967.1s+2467s^{0.5}}\end{bmatrix}. \tag{5.4}$$

Let us consider $\theta(s)$ for further analysis:

$$\theta(s) = K\frac{480s + 164470}{s^{0.5}(s^{1.5} + 7.2s + 967.1s^{0.5} + 2467)}, \tag{5.5}$$

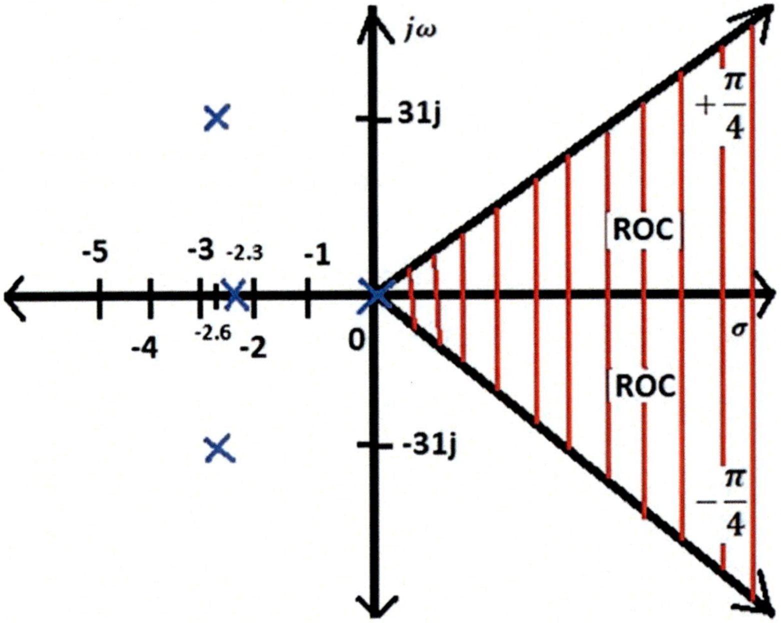

Fig. 5.3 Region of convergence of $\theta(s)$

Through partial fraction expansion, we get

$$\frac{\theta(s)}{K} = \frac{65.46}{s^{0.5}} - \frac{67.12}{s^{0.5}+2.6} + \frac{0.83-5j}{s^{0.5}+2.3-31j} + \frac{0.83+5j}{s^{0.5}+2.3+31j}. \tag{5.6}$$

Before determining the inverse Laplace transform of (5.6), let us check the ROC of the system. The ROC of a stable system represented by (5.6) is obtained as $Re\{s^{0.5}\} > 0$, $Re\{s^{0.5}\} > -2.6$ and $Re\{s^{0.5}\} > -2.3$. All ROCs contain the wedge shape axis. Hence, the overall ROC is the intersection of all these ROCs: $Re\{s^{0.5}\} > 0 \cap Re\{s^{0.5}\} > -2.6 \cap Re\{s^{0.5}\} > -2.3 = Re\{s^{0.5}\} > 0$.

Figure 5.3 shows the ROC of (5.6). Taking the inverse Laplace transform for stable $\theta(s)$ to convert (5.6) into time domain using definitions of Mittag-Leffler (ML) function [46], we get

$$\begin{aligned}\frac{\theta(t)}{K} &= \frac{65.46t^{-0.5}}{\Gamma(0.5)} - 65.32t^{-0.5}E_{0.5,0.5}(-2.6t^{0.5}) \\ &+ (0.175+2.7j)t^{-0.5}E_{0.5,0.5}(-(2.3-31j)t^{0.5}) \\ &- (0.173+2.7j)t^{-0.5}E_{0.5,0.5}(-(2.3+31j)t^{0.5}).\end{aligned} \tag{5.7}$$

By similar steps, it is possible to determine the inverse Laplace transform pertaining to $\phi(t)$ as well:

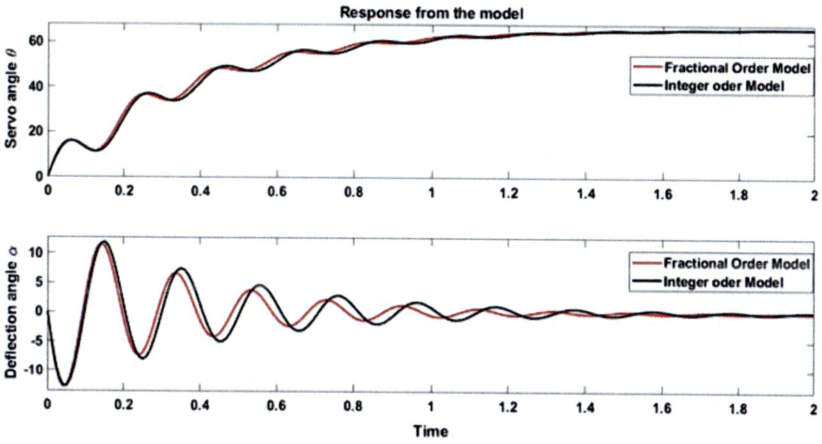

Fig. 5.4 Model validation for servo angle

$$
\begin{aligned}
\frac{\phi(t)}{K} &= 1.3t^{-0.5}E_{0.5,0.5}(-2.6t^{0.5}) \\
&- (0.65 + 7.7j)t^{-0.5}E_{0.5,0.5}(-(2.3 - 31j)t^{0.5}) \\
&- (0.65 + 7.7j)t^{-0.5}E_{0.5,0.5}(-(2.3 + 31j)t^{0.5}).
\end{aligned}
\tag{5.8}
$$

To validate the model obtained in (5.7) and (5.8), an impulse input is provided to (5.3) and also to (5.7), (5.8) and the responses obtained are compared as shown in Fig. 5.4. To obtain the results of Fig. 5.4, K is found to be 1.025 by hit and trail method such that the fractional behaviour of the considered manipulator works in the same manner as the integer behaviour of this manipulator, and for simplicity α is chosen to be 0.5.

From Fig. 5.4, it can be concluded that the FO model of SFLRM tracks the integer (IO) model, and hence the fractional model (5.7) and (5.8) of SFLRM is valid.

5.2 Stability Analysis and Control Law Design

In this section, a stability analysis is carried in Ω-Domain [17]. Let us use the transformation from s-domain to Ω-domain using $s^\alpha = \Omega$. The Ω- domain looks like Fig. 5.5. Using the Ω-domain transformation, we study the Ω-plane poles. Once the time domain responses of a system are obtained corresponding to the Ω-plane pole locations, the behavior of such systems in the new complex plane can be characterized:

$$
\Omega = s^\alpha = (re^{j\theta})^\alpha = r^\alpha e^{j\theta\alpha}. \tag{5.9}
$$

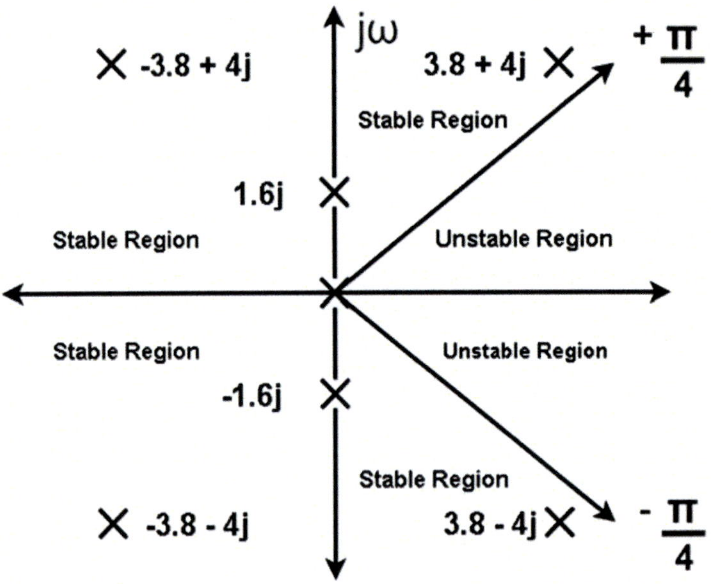

Fig. 5.5 Ω-domain analysis of the poles of the SFLRM system

Using (5.9), it is possible to map s-plane to Ω-plane. For stability, a mapping of imaginary axis is important, i.e, $s = re^{\pm\frac{j\pi}{2}}$. Image of this axis in Ω-plane is represented as

$$\Omega = r^{\alpha}e^{\pm\frac{j\alpha\pi}{2}}, \tag{5.10}$$

which represents a pair of lines at $\gamma = \pm\frac{\alpha\pi}{2}$ where γ is angle in Ω-plane and $\Omega = \rho e^{j\gamma}$. The right half of s-plane maps into wedge shape in the Ω-plane. In Fig. 5.5, whenever the system poles lie to the left of $\pm\alpha\frac{\pi}{2}$ axis in Ω-domain, then the system is said to stable, else unstable.

Considering α in (5.3) to be 0.5 (for simplicity) and using the $s^{0.5} = \Omega$ transformation, the locations of the poles are found to be 0, 0, $3.8 \pm 4j$, $-3.8 \pm 4j$ and $\pm 1.6j$. The locations of these poles are also shown in Fig. 5.5. The SFLRM system is stable.

Let us take fractional model (5.4) and choose $\alpha = 0.5$ (for simplicity) to design control law. The state-space form can be modified and will look like:

$$\begin{aligned} {}_0D_t^{\,0.5}x &= A_t x + B_t u, \\ y &= C_t x, \end{aligned} \tag{5.11}$$

where

$$A_t = \begin{bmatrix} 0 & -7.2 & 0 & -967.1 & 0 & -2467 & 0 & 0 \\ 1 & 0 & 0 & 0 & 0 & 0 & 0 & 0 \\ 0 & 1 & 0 & 0 & 0 & 0 & 0 & 0 \\ 0 & 0 & 1 & 0 & 0 & 0 & 0 & 0 \\ 0 & 0 & 0 & 1 & 0 & 0 & 0 & 0 \\ 0 & 0 & 0 & 0 & 1 & 0 & 0 & 0 \\ 0 & 0 & 0 & 0 & 0 & 1 & 0 & 0 \\ 0 & 0 & 0 & 0 & 0 & 0 & 1 & 0 \end{bmatrix},$$

$$B_t = \begin{bmatrix} 1 & 0 & 0 & 0 & 0 & 0 & 0 & 0 \end{bmatrix}^T, \quad C_t = \begin{bmatrix} 0 & 0 & 0 & 480 & 0 & 0 & 0 & 164470 \\ 0 & 0 & 0 & 480 & 0 & 0 & 0 & 0 \end{bmatrix}.$$

It is required to check the controllability of the above system before designing the control law. The controllability matrix (CM) is given as

$$CM = \begin{bmatrix} 1 & 0 & 7 & 0 & 915 & 0 & 11086 & 0 \\ 0 & 1 & 0 & 7 & 0 & 915 & 0 & 11086 \\ 0 & 0 & 1 & 0 & 7 & 0 & 915 & 0 \\ 0 & 0 & 0 & 1 & 0 & 7 & 0 & 915 \\ 0 & 0 & 0 & 0 & 1 & 0 & 7 & 0 \\ 0 & 0 & 0 & 0 & 0 & 1 & 0 & 7 \\ 0 & 0 & 0 & 0 & 0 & 0 & 1 & 0 \\ 0 & 0 & 0 & 0 & 0 & 0 & 0 & 1 \end{bmatrix}.$$

It is clear that the system matrix is full rank and hence controllable. For pole placement control [48, 49], let us formulate a feedback control law given by $u = -\mathbf{k}x$, a state feedback form. The pole placement design begins with an arbitrary assumption of what form the controller must take in order to control the given plant, and a symbolic characteristic equation is formulated. Then the desired closed-loop poles are provided based on design attributes like overshoot, settling time, etc. Most of the time the final characteristic equation has more than 2 poles. Once the closed loop poles are decided, a desired characteristic equation is formed. The coefficients for each power of s are equated from the symbolic characteristic equation to the desired. Algebra is used to determine the controller coefficients necessary to achieve the desired closed-loop poles with the assumed controller form.

From (5.11), the closed-loop system is given as

$$_0D_t^{0.5}x = (A_t - B_t k)x = A_{CL}x. \tag{5.12}$$

Next, the aim is to find the gain matrix $\mathbf{k}$ such that the system gives the desired characteristics. Matrix A_t is in first companion (i.e. controllable canonical) form. The gain matrix $\mathbf{k}$ is designed in such a way that satisfies

$$|\Omega I - (A_t - B_t k)| = (\Omega - \mu_1)(\Omega - \mu_2)(\Omega - \mu_3).....(\Omega - \mu_n), \tag{5.13}$$

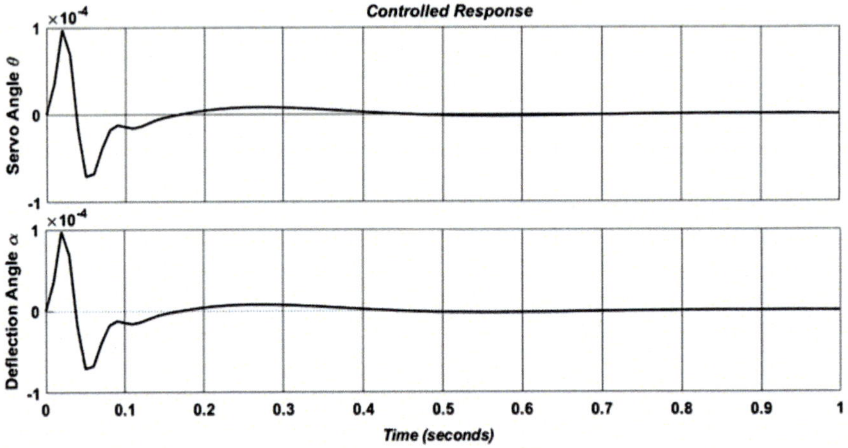

Fig. 5.6 Error output of the SFLRM system by pole placement method

where μ is the desired pole locations for the considered system. Using (5.13) solve for **k** by equating the like powers on both side. Taking only the terms on the left hand side of (5.14), we get

$$|\Omega I-(A_t-B_tk)|=\Omega^8-k_1\Omega^7-(k_2-7.2)\Omega^6-(k_4-967.1)\Omega^4 \\ -k_3\Omega^5-k_5\Omega^3-(k_6-2467)\Omega^2-k_7\Omega-k_8. \quad (5.14)$$

Let us have a constraint that our system has to provide stable response before 1.5 sec with minimum overshoot. Then the desired poles of the system are to be placed at $-50+100i, -50-100i, -5-10i, -5+10i, -80+100i, -80-100i, -5+1.6071i, -5-1.6071i$. The terms on the right hand side of (5.13) can be expressed as

$$\Omega^8+280\Omega^7+5.035\times10^4\Omega^6+4.605\times10^6\Omega^5+2.895\times10^8\Omega^4 \\ +5.089\times10^9\Omega^3+5.749\times10^{10}\Omega^2+3.253\times10^{11}\Omega+7.068\times10^{11}. \quad (5.15)$$

On comparing the like powers of (5.14) and (5.15), we can have $\mathbf{k}=[-280, -5\times10^4, -4.6\times10^6, -2.8\times10^8, -5\times10^9, -5.7\times10^{10}, -3.2\times10^{11}, -7\times10^{11}]$, and hence the control law can be expressed as

$$u=280x_1+5\times10^4x_2+4.6\times10^6x_3+2.8\times10^8x_4 \\ +5\times10^9x_5+5.7\times10^{10}x_6+3.2\times10^{11}x_7+7\times10^{11}x_8. \quad (5.16)$$

The response of this system is shown in Fig. 5.6.

From Fig. 5.6, it can be noticed that the system reaches steady state before the desired set time of 1.5 s. Hence, the aim of designing the control law for SFLRM for an error-free response in minimum time, is achieved. Next, a real-time single robotic manipulator is considered to validate the control law.

5.3 Experimental Validation

Experimental validation of the proposed FO control algorithms applied on FO dynamical models, is needed to establish the efficacy of the proposed methodology. In this book, three experimental setups have been used: (i) A real-time single link robotic manipulator, (ii) 2DOF Serial Flexible Link Robotic Manipulator, and (iii) 2DOF Serial Flexible Joint Robotic Manipulator. Of these, the last two are considered in the next chapter. The considered robotic manipulator is a single flexible link robotic manipulator, a piece of educational equipment from Quanser, to check the concepts developed in this paper. The device has a rotary system to rotate the flexible arm of the robotic manipulator to the desired position as shown in Fig. 5.7. This manipulator has a flexible arm, and this link (arm) can be positioned anywhere in between $-45°$ to $45°$ or as per the desired requirement. The controller is designed using MATLAB 2018b version on the Intel i3 CPU with 4GB RAM to control the angular position θ of this manipulator.

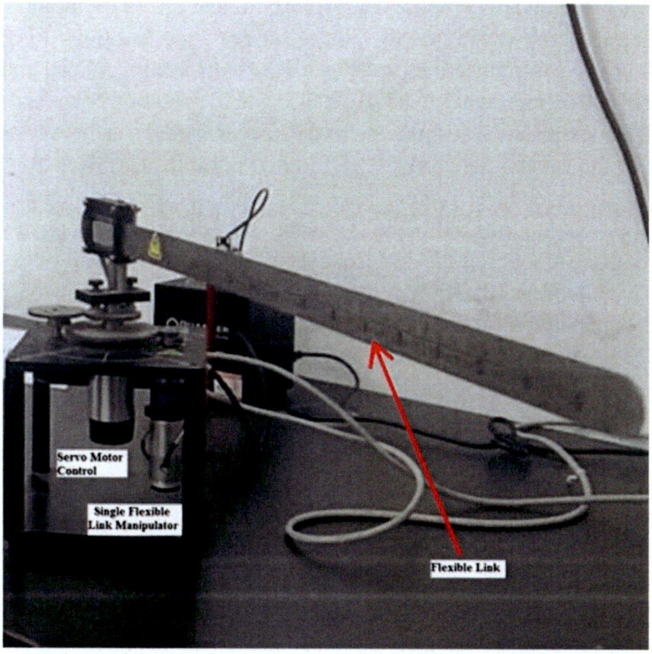

Fig. 5.7 Experimental setup of single flexible link robotic manipulator

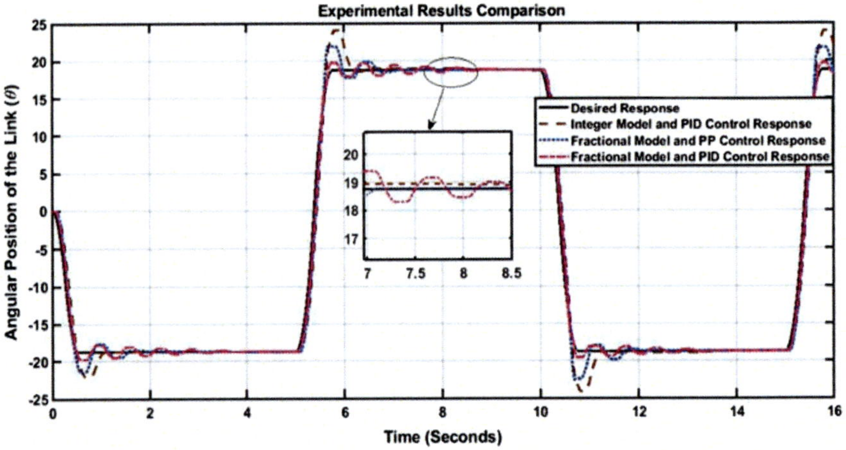

Fig. 5.8 Controlled output response of the SFLRM system

The experimental response of the controlled output is shown in Fig. 5.8. The considered robotic manipulator is supposed to track a square wave pulse. The figure shows the experimental output response of a PID controller designed for integer model and fractional model of the considered robotic system. The PID controller response is presented to compare the results obtained by the PP method of controller design. It can be observed that the controller designed using the PP method has a better control performance compared to PID controller for integer model and PID controller for the fractional model. By looking at the zoomed portion, one can notice that the PID controller corresponding to the integer model has the worst performance in terms of the steady-state error. The zoomed portions also show that when a PID controller is designed for fractional system model, the system may have oscillatory behavior around the desired position. If the control law is designed using the PP method for the fractional model, the system performance improves significantly with no overshoot and the system has a minimum settling time.

Another experimental work is undertaken on the single link robotic manipulator. The integer model and the fractional model of this manipulator are considered for the design of fractional PID and LQR control, as shown in Fig. 5.9. The control responses are studied in comparison to the responses obtained through the pole placement control design. The response of the PID controller for the angular position of integer model of the considered robotic manipulator settles in 3.4 s. The LQR control of the angular position of the integer model of the robotic manipulator settles in 3.2 s. The LQR control of the angular position of the fractional model of the considered robotic manipulator settles in 2.2 s. From Fig. 5.8 it is observable that the pole placement control strategy makes the angular position of the considered robotic manipulator stable in 1.8 s. So, from this analysis and from the responses, it can be observed that a control strategy designed using pole placement method has a better settling time for the robotic manipulator.

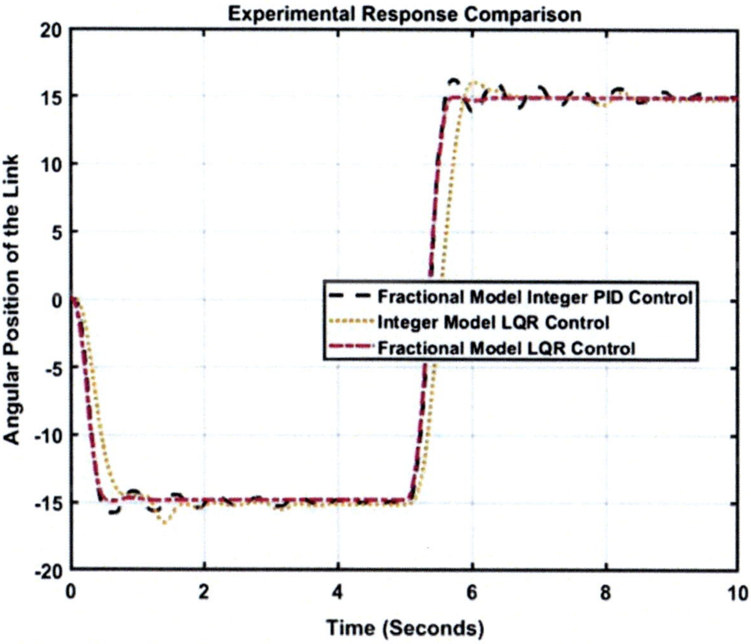

Fig. 5.9 PID and LQR controlled output response of the SFLRM system

Table 5.1 Controlled response analysis of SFLRM

Specifications⟶ / Controller↓	Settling time (sec)	% Improvement in settling time using PP control
Integer model and PID control	3.4	≈89%
Fractional model and PID control	3	≈67%
Integer model and LQR control	3.2	≈78%
Fractional model and LQR control	2.2	≈22%
Fractional model and PP control	1.8	–

Table 5.1 gives a statistical analysis of the control methods used in this paper for fractional-order and integer-order model of the SFLRM. The table shows that a PID controller for the integer-order SFLRM model takes 3.4 sec to track the desired response. A PID controller for the fractional-order SFLRM model takes 3 sec to track the desired response, LQR control for integer model takes 3.2 sec to track the desired response and LQR control for fractional model takes 2.2 sec to track the desired response. But, a controller designed using the pole-placement (PP) method for the fractional-order model of this system, takes 1.8 sec to track the desired response.

That is, there is an 89% and 22% improvement in the settling time when compared to a controller for integer model, 78% and 67% improvement in the settling time compared to a controller for the fractional model. Hence, this system's response is improved significantly with a PP method of controller design.

In this chapter, a single flexible link robotic manipulator system has been considered for modeling, stability analysis, and control law design using fractional calculus representation. The fractional modeling of the system is reported using the concept of fractional order Laplace Transform. The model is then validated and it is found that the fractional model of SFLRM system presented in this paper follows the properties of the integer-order SFLRM system. The fractional stability analysis of this model is carried out in the Ω-domain and it is found that all the poles of this system lie on the left of the Ω transformed plane, and hence the SFLRM robotic manipulator system is stable. The control law of the fractional SFLRM system is derived using pole placement method. The aim to design a controller which makes the system stable with minimum overshoot and settling time is achieved in this paper. The control law obtained by PP is experimentally validated on a real-time setup of single link robotic manipulator and the response is also compared with PID controller for the same system. It is found that when the controller is designed for single link robotic manipulator using pole placement method, the system performance is better. The concept of fractional model and controller design proves to be better performing. This methodology of fractional modeling and controller design is not limited to controlling the robotic manipulators, but also in the other areas where more accurate system models are required such as process control where one can find a fractional equivalent model of a chemical reaction, in the field of water pH monitoring, in the area of a temperature control application and so on.

References

1. Sethi, J.K., Deb, D., Malakar. M.: Modeling of a wind turbine farm in presence of wake interactions. In: 2011 International Conference On Energy, Automation And Signal (2011)
2. Patel, R., Deb, D.: Parametrized control-oriented mathematical model and adaptive backstepping control of a single chamber single population microbial fuel cell. J. Power Sources **396**, 599–605 (2018)
3. Nath, A., Deb, D., Dey, R.: An augmented subcutaneous type 1 diabetic patient modelling and design of adaptive glucose control. J. Process Control **86**, 94–105 (2020)
4. Nasser-Eddine, A., Huard, B., Gabano, J.D., Poinot, T.: A two steps method for electrochemical impedance modeling using fractional order system in time and frequency domains. Control Eng. Pract. **86**, 96–104 (2019)
5. Qureshi, S., Yusuf, A., Shaikh, A.A., Inc, M., Baleanu, D.: Fractional modeling of blood ethanol concentration system with real data application. Chaos Interdiscip. J. Nonlinear Sci. **29**(1), 013143 (2019)
6. Machado, J.T., Lopes, A.M.: Fractional-order modeling of a diode. Commun. Nonlinear Sci. Numer. Simul. **70**, 343–353 (2019)
7. Singh, A.P., Deb, D., Agarwal, H.: On selection of improved fractional model and control of different systems with experimental validation. Commun. Nonlinear Sci. Numer. Simul. **79**, 104902 (2019)

8. Singh, A.P., Agarwal, H., Srivastava, P.: Fractional order controller design for inverted pendulum on a cart system (POAC). WSEAS Trans. Syst. Control **10**, 172–178 (2015)
9. Singh, A., Agrawal, H.: A fractional model predictive control design for 2-d gantry crane system. J. Eng. Sci. Technol. **13**(7), 2224–2235 (2018)
10. Mujumdar, A., Tamhane, B., Kurode, S.: Fractional order modeling and control of a flexible manipulator using sliding modes. In: 2014 American Control Conference. IEEE, pp. 2011–2016 (2014)
11. Kexue, L., Jigen, P.: Laplace transform and fractional differential equations. Appl. Math. Lett. **24**(12), 2019–2023 (2011)
12. Lin, S.D., Lu, C.H.: Laplace transform for solving some families of fractional differential equations and its applications. Adv. Differ. Equ. **2013**(1), 137 (2013)
13. Sabatier, J., Farges, C., Trigeassou, J.C.: Fractional systems state space description: some wrong ideas and proposed solutions. J. Vib. Control **20**(7), 1076–1084 (2014)
14. Li, C.P., Zhang, F.R.: A survey on the stability of fractional differential equations. Eur. Phys. J. Spec. Top. **193**(1), 27–47 (2011)
15. Li, Y., Chen, Y., Podlubny, I.: Mittag-Leffler stability of fractional order nonlinear dynamic systems. Automatica **45**(8), 1965–1969 (2009)
16. Tavazoei, M.S., Haeri, M.: A note on the stability of fractional order systems. Math. Comput. Simul. **79**(5), 1566–1576 (2009)
17. Bandyopadhyay, B., Kamal, S.: Solution, stability and realization of fractional order differential equation. In: Stabilization and Control of Fractional Order Systems: A Sliding Mode Approach. Springer, Cham, pp. 55–90 (2015)
18. Abdulwahhab, O.W., Abbas, N.H.: A new method to tune a fractional-order PID controller for a twin rotor aerodynamic system. Arab. J. Sci. Eng. **42**(12), 5179–5189 (2017)
19. Dabiri, A., Poursina, M., Butcher, E.A.: Integration of divide-and-conquer algorithm with fractional order controllers for the efficient dynamic modeling and control of multibody systems. In: 2018 Annual American Control Conference (ACC). IEEE, pp. 4201–4206 (2018)
20. Copot, C., Muresan, C.I., Markowski, K.A.: Advances in fractional order controller design and applications. J. Appl. Nonlinear Dyn. **8**(1), 1–3 (2019)
21. Bingi, K., Ibrahim, R., Karsiti, M.N., Hassan, S.M.: Fractional order set-point weighted PID controller for pH neutralization process using accelerated PSO algorithm. Arab. J. Sci. Eng. pp. 1–15 (2017)
22. Bingi, K., Ibrahim, R., Karsiti, M.N., Hassan, S.M., Harindran, V.R.: Real-time control of pressure plant using 2DOF fractional-order PID controller. Arab. J. Sci. Eng. pp. 1–12 (2018)
23. Pandey, S., Dwivedi, P., Junghare, A.: Anti-windup fractional order $PI^{\lambda} - PD^{\mu}$ controller design for unstable process: a magnetic levitation study case under actuator saturation. Arab. J. Sci. Eng. **42**(12), 5015–5029 (2017)
24. Pandey, S., Dwivedi, P., Junghare, A.S.: A newborn hybrid anti-windup scheme for fractional order proportional integral controller. Arab. J. Sci. Eng. **43**(6), 3049–3063 (2018)
25. Kumar, A., Kumar, V.: Design of interval type-2 fractional-order fuzzy logic controller for redundant robot with artificial bee colony. Arab. J. Sci. Eng. pp. 1–20 (2018)
26. Kerrouche, K.D.E., Wang, L., Mezouar, A., Boumediene, L., Van Den Bossche, A.: Fractional-order sliding mode control for D-STATCOM connected wind farm based DFIG under voltage unbalanced. Arab. J. Sci. Eng. pp. 1–16 (2018)
27. Dabiri, A., Butcher, E.A.: Optimal observer-based feedback control for linear fractional-order systems with periodic coefficients. J. Vib. Control **25**(7), 1379–1392 (2019)
28. Dabiri, A., Butcher, E.A., Poursina, M., Nazari, M.: Optimal periodic-gain fractional delayed state feedback control for linear fractional periodic time-delayed systems. IEEE Trans. Autom. Control **63**(4), 989–1002 (2017)
29. Gong, Y., Wen, G., Peng, Z., Huang, T., Chen, Y.: Observer-based time-varying formation control of fractional-order multi-agent systems with general linear dynamics. In: IEEE Transactions on Circuits and Systems II: Express Briefs (2019)

30. Cortez, A.J.G., Mendez-Barrios, C.F., González-Galván, E.J., Mejía-Rodríguez, G., Félix, L.: Geometrical design of fractional PD controllers for linear time-invariant fractional-order systems with time delay. Proc. Inst. Mech. Eng. Part I: J. Syst. Control Eng. **233**(7), 815–829 (2019)
31. Boubellouta, A., Boulkroune, A.: Intelligent fractional-order control-based projective synchronization for chaotic optical systems. Soft Comput. **23**(14), 5367–5384 (2019)
32. Munoz-Hernandez, G.A., Mino-Aguilar, G., Guerrero-Castellanos, J.F., Peralta-Sanchez, E.: Fractional order PI-based control applied to the traction system of an electric vehicle (EV). Appl. Sci. **10**(1), 364 (2020)
33. Birs, I., Muresan, C., Nascu, I., Ionescu, C.: A survey of recent advances in fractional order control for time delay systems. IEEE Access **7**, 30951–30965 (2019)
34. Sanz, A., Etxebarria, V.: Composite robust control of a laboratory flexible manipulator. In: Proceedings of the 44th IEEE Conference on Decision and Control. IEEE, pp. 3614–3619 (2005)
35. Etxebarria, V., Sanz, A., Lizarraga, I.: Control of a lightweight flexible robotic arm using sliding modes. Int. J. Adv. Robot. Syst. **2**(2), 11 (2005)
36. Mujumdar, A.A., Kurode, S.: Second order sliding mode control for single link flexible manipulator. In: International Conference on Machines and Mechanisms (2013)
37. Shitole, C., Sumathi, P.: Sliding DFT-based vibration mode estimator for single-link flexible manipulator. IEEE/ASME Trans. Mechatron. **20**(6), 3249–3256 (2015)
38. Mujumdar, A., Tamhane, B., Kurode, S.: Observer-based sliding mode control for a class of noncommensurate fractional-order systems. IEEE/ASME Trans. Mechatron. **20**(5), 2504–2512 (2015)
39. Ahmad, M.A., Mohamed, Z., Ismail, Z.H.: Hybrid input shaping and PID control of a flexible robot manipulator. J. Inst. Eng. **72**(3), 56–62 (2009)
40. Pham, D.T., Koç, E., Kalyoncu, M., Tınkır, M.: Hierarchical PID controller design for a flexible link robot manipulator using the bees algorithm. Methods (eg. genetic algorithm), 25, 32 (2008)
41. Jnifene, A., Andrews, W.: Experimental study on active vibration control of a single-link flexible manipulator using tools of fuzzy logic and neural networks. IEEE Trans. Instrum. Meas. **54**(3), 1200–1208 (2005)
42. Sun, C., He, W., Hong, J.: Neural network control of a flexible robotic manipulator using the lumped spring-mass model. IEEE Trans. Syst. Man Cybern. Syst. **47**(8), 1863–1874 (2016)
43. Sun, C., Gao, H., He, W., Yu, Y.: Fuzzy neur al network control of a flexible robotic manipulator using assumed mode method. IEEE Trans. Neural Netw. Learn. Syst. **99**, 1–14 (2018)
44. Forbes, J.R., Damaren, C.J.: Single-link flexible manipulator control accommodating passivity violations: theory and experiments. IEEE Trans. Control Syst. Technol. **20**(3), 652–662 (2011)
45. Talole, S.E., Kolhe, J.P., Phadke, S.B.: Extended-state-observer-based control of flexible-joint system with experimental validation. IEEE Trans. Ind. Electron. **57**(4), 1411–1419 (2009)
46. Haubold, H.J., Mathai, A.M., Saxena, R.K.: Mittag-Leffler functions and their applications. In: Journal of Applied Mathematics (2011)
47. Quanser Inc. SRV02 Rotary Flexible Link User Manual (2011)
48. Deb, D., Tao, G., Burkholder, J., Smith, D.: An adaptive inverse control scheme for a synthetic jet actuator model. In: Proceedings Of The 2005, American Control Conference (2005). https://doi.org/10.1109/acc.2005.1470367
49. Deb, D., Tao, G., Burkholder, J., Smith, D.: An adaptive inverse control scheme for synthetic jet actuator arrays. In: Infotech@Aerospace (2005). https://doi.org/10.2514/6.2005-7170

Chapter 6
Improved Fractional Model Selection and Control with Experimental Validation

Under certain conditions, approximation based control performs effectively [1]. Most of the dynamic systems are better characterized using a fractional-order (FO) dynamic model based on fractional calculus [2]. When dealing with FO systems, a major challenge is to select the fractional values that generally termed as α. Many researchers have proposed different medium to choose appropriate α values whether it belongs to the fractional model or fractional controller. A ranking based on various researchers working in the field of fractional calculus is reported in [3]. Monje et al. delineate an idea of tuning the fractional value for FO PI^{α} controller [4] and recommend tuning of the controller using an iterative optimization method based on a nonlinear function minimization. In [5], fractional order controller is designed for FO plant which supervises the heating furnace system to a general modified FO model and PID controller is designed. In [6], a method for tuning fractional $PI^{\lambda}D^{\mu}$ controller is proposed to fulfil five different design specifications, and in [7], a new tuning method for FO proportional and the derivative controller is proposed for a class of second-order plants, and the plant model is an integer order model. In [8], fractional order controller is designed for a class of FO systems. In article [9] a set of tuning rules is presented for FO controllers based on a first-order-plus-dead-time model. A FO sliding mode control is presented in [10] to control the velocity of the permanent magnet synchronous motor. In [11], a tuning method of FO controller is presented for a class of FO systems. Recently in [12], a FOPID controller is designed for FO systems. A new practical tuning method development for FOPI controller is presented in [13] which is also valid for a general class of plants, and another algorithm stabilizes FO time-delay systems using FO PID controllers [14].

There are also other studies which have derived FO models and its control [15–17]. In [18], authors reported different integer and fractional order models from the electrical point of view. Modeling of aging materials within the limitation of fractional calculus consisting of variable order is presented [19]. A FO methodology of capacitor quality control for porous electrode behavior is observed [20]. The

A. P. Singh et al., *Fractional Modeling and Controller Design of Robotic Manipulators*, Intelligent Systems Reference Library 194,
https://doi.org/10.1007/978-3-030-58247-0_6

advantage of such FO modeling is the absence of high-order polynomials in transfer functions. In [21], a detailed tutorial is presented for the representation of a FO system in MATLAB. More recently, a study of fractional modeling and fractional controller design for an inverted pendulum system has been provided [22]. An intuitive study of Variable order fractional systems with an inverse Laplace transform is presented [23]. A fraction model of semiconductor PN junction diode is proposed [24]. A detailed review of literature related to fractional derivatives is also presented [25]. An interesting to describe the dynamics of political systems is presented [26].

Two robotic manipulators are considered: serial flexible link robotic manipulator, and serial flexible joint robotic manipulator, both with two degrees of freedom. Their fractional equivalent model is derived using the proposed algorithm, and then a controller is designed for three different scenarios, i.e., fractional model and integer Controller, integer model and fractional controller, fractional model and fractional controller. The simulated results are compared with the experimentally examined results and it is found that the proposed fractional model of these robotic manipulators is highly accurate and reliable. The examined results show significantly improved performance using the fractional controller and fractional model and strong evidence of the reliability of the fractional model in control system design. Various efforts have taken place to tune/find the appropriate α value. Most of the papers above have taken a system model (which is integer or fractional) and tuned a fractional PID controller to get the desired response. There is no unique way to choose the appropriate value of α (for a system's fractional model). Fractional Calculus is a powerful new tool that models different dynamical systems whose modeling equations are in the form of differential equation to study the fractional dynamical behavior (accurate behavior). The modeling equations of the dynamical systems which are taken as an example and for experimentally validating the results are represented as differential equations. Hence, fractional calculus is chosen for further investigation applied to robotic manipulators. This chapter contributes to the area of choosing an appropriate value of 'α', detecting the best fractional model and designing a controller for the derived fractional model. The contributions of this chapter are summarized as

1. *An algorithm to select the best suitable fractional model is proposed.*
2. *Proposed an improved fractional model of a system.* Different sets of second-order systems are considered, and the corresponding suitable fractional model is reported. The performance merit of the algorithm is evaluated against some pre-existing models via a simulation experiment. The FO model obtained by the proposed algorithm performs significantly better. Improvement of 85% in settling time is achieved against example 2 of [6] with zero overshoot and an improvement of 50.2%, 93.8% respectively are achieved against the FO model of heating furnace [27].
3. Proposed fractional model of two robotic manipulators of two degrees of freedom: (i) serial flexible link, and (ii) serial flexible joint.
4. *Experimental Contribution.* Simulated results are compared with practical examined results and the proposed fractional model of these robotic manipulators are proven to be highly accurate and reliable.

6.1 Algorithm for Selection of a Fractional Model

A pseudo-code of the algorithm is presented below. An integer order transfer function is applied to this algorithm and is converted to its equivalent FO transfer function using the definition of fractional order Laplace transformation [16, 25]. Fractional equivalent transfer function is approximated [17] for each α values between 0 to 1. All the controlled responses are compared, and the one which gives the best result is then chosen. When the desired specification is met, the algorithm displays the best FO model.

Algorithm for Fractional Value Selection

Input: Transfer function, α initial guess, Simulink Controller
Output: Best Fractional Order α Value, Best Settling Time Ts

```
Define a Linear Transfer Function eg. 1/s
i ← 1
α[i] ← 0.1
for i = 1 : 10 do
  Derive the Fractional Transfer Function 1/s^α[i]
  Approximate The Transfer Function
  Find the Response of System
  Store the Settling Time Ts[i]
  Store the α[i]
  α[i + 1] = α[i] + 0.1
  i = i + 1
end for
minimum1 = Ts(1)
α_location = 1
for ii = 2 : numel(Ts) do
  if Ts(ii) < minimum1 then
    minimum1 = Ts(ii)
    α_location = ii
  else
    ii = ii + 1
  end if
  Display Ts = minimum1
  Display α = α(α_location)
end for
```

Let us consider a second order system

$$TF(s) = \frac{\omega_n}{s^2 + 2\zeta\omega_n s + {\omega_n}^2}, \tag{6.1}$$

assuming $\omega_n = 50$ rad/sec, $\zeta = 1$. Mapping the transfer function obtained in (6.1) to its fractional order equivalent transfer function, we get

$$TF(s) = \frac{50}{s^{2\alpha} + 100s^{\alpha} + 50^2}, \tag{6.2}$$

where α is the fractional value. The aim is to choose the best fractional order model depending on α, corresponding to (6.1) to design a PID controller that can give better settling time by using the proposed algorithm. Using the transformation [26, 28]

$$\Omega = s^{\alpha} = (re^{j\theta})^{\alpha} = r^{\alpha}e^{j\theta\alpha}, \tag{6.3}$$

it is possible to map s-plane to Ω-plane as represented by (6.3). For stability, a mapping of imaginary axis is important i.e. $s = re^{\pm\frac{j\pi}{2}}$. Image of this axis in Ω-plane is represented by

$$\Omega = r^{\alpha}e^{\pm\frac{j\alpha\pi}{2}}, \tag{6.4}$$

which represents a pair of lines at $\gamma = \pm\frac{\alpha\pi}{2}$ where γ is angle in Ω-plane and $\Omega = \rho e^{j\gamma}$. The right half of s-plane maps into a wedge shape in the Ω-plane. When the system poles lie left of $\pm\alpha\frac{\pi}{2}$ axis in Ω-domain, then that system is said to be stable, else unstable. Considering α in (6.2) to be 0.5 (for simplicity) and using $s^{0.5} = \Omega$ transformation. The locations of the poles are found to be $8 \times 10^{-11} \pm 7.07j$ and $-8 \times 10^{-11} \pm 7.07j$. The locations of these poles are shown in Fig. 6.1 representing a stable second-order system.

Now, let us obtain the appropriate α value of (6.2) using the proposed algorithm proposed which takes less than a minute to select the best fractional model for one decimal place of α. The same algorithm takes 10–15 min to detect the best fractional

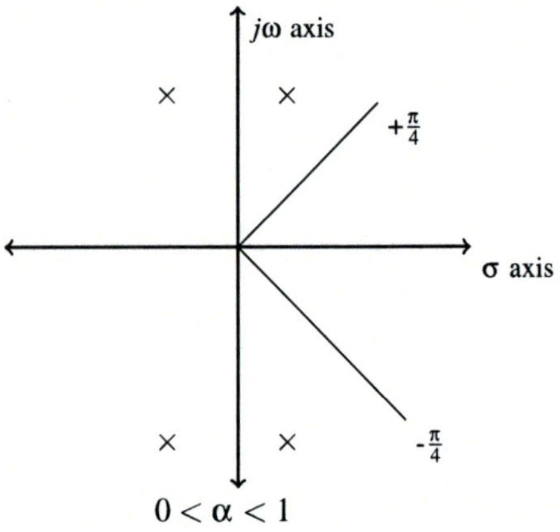

Fig. 6.1 Ω-domain analysis of the poles of a second order system. All the poles are in the left of the wedge shaped region and hence the system is stable

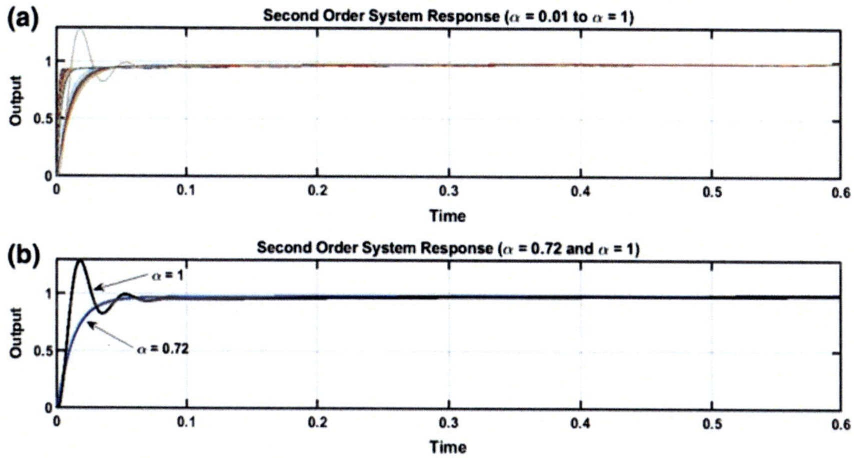

Fig. 6.2 PID Controller response of second order system **a** For α values between 0.01 to 1, **b** for $\alpha = 0.72$ (appropriate fractional order system) and $\alpha = 1$ (integer order system)

model for two decimal places of α value and around 60–70 min to detect the best fractional model corresponding to the third decimal place of α value. This algorithm may take hours depending on the α values if the decimal places are further increased. A PID controller is taken into the loop of the algorithm, and the controller response of the second-order systems is shown in Fig. 6.2.

By eliminating all the other PID controller responses from Fig. 6.2a, which do not provide the desired output except for $\alpha = 1$ and $\alpha = 0.72$ (the best model) we get Fig. 6.2b. The settling time corresponding to this fractional model is 0.3 sec with zero overshoot, which is less and far better as compared to the settling time of the integer second order model which is 0.7 sec and overshoot of 29%. The fractional equivalent transfer function model of (6.2) using the proposed algorithm is given as

$$TF(s) = \frac{50}{s^{1.44} + 100s^{0.72} + 2500}, \tag{6.5}$$

and the approximate equivalent transfer function is found to be

$$G(s) = \frac{0.1223s}{s^2 + 6.119s + 0.00002}. \tag{6.6}$$

6.2 Evaluation of the Algorithm

Assuming mass m to be 1 kg, damping constant b to be 10 Ns/m and spring constant k to be 20 N/m, then the transfer function corresponding to the MSD system is

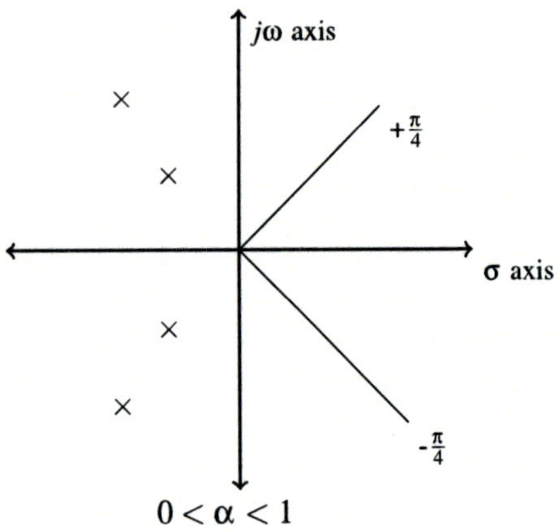

Fig. 6.3 Ω-domain analysis of the poles of a MSD system, all of which are in the left of the wedge shaped region, thereby ensuring stability

$$G(s) = \frac{1}{s^2 + 10s + 20}. \tag{6.7}$$

Now, let us convert (6.7) into its fractional equivalent transfer function

$$G(s) = \frac{1}{s^{2\alpha} + 10s^{\alpha} + 20}, \tag{6.8}$$

where α is the fractional value. Using the transformation (6.3) and Considering α in (6.8) to be 0.5 for simplicity, the locations of the poles are found to be $-6 \times 10^{-17} \pm 2.69j$ and $-2.2 \times 10^{-17} \pm 1.66j$ and are shown in Fig. 6.3. All the poles are lying left to the wedge shaped unstable region and the considered MSD system is a stable.

Using two decimal places, the algorithm gives a more accurate fractional model of the considered system. When the transfer function (6.8) is provided to the proposed algorithm for $\alpha = 0.01$ to $\alpha = 1$, the resultant PID controller responses are shown in Fig. 6.4. Let us eliminate all the other PID controller responses from Fig. 6.4a that do not provide the desired output except for $\alpha = 1$ and $\alpha = 0.32$ (the best model) in Fig. 6.4b. The settling time corresponding to this fractional model is 0.8 sec with 0.5762% overshoot, which is less and far better as compared to the settling time of the integer order model 1.4 sec and overshoot of 9.8404%. The fractional equivalent transfer function of this model is found to be

$$TF(s) = \frac{1}{s^{0.64} + 10s^{0.32} + 20}, \tag{6.9}$$

and the approximate equivalent transfer function is found to be

$$G(s) = \frac{0.1398s + 0.001637}{s^2 + 3.338s + 0.03456}. \tag{6.10}$$

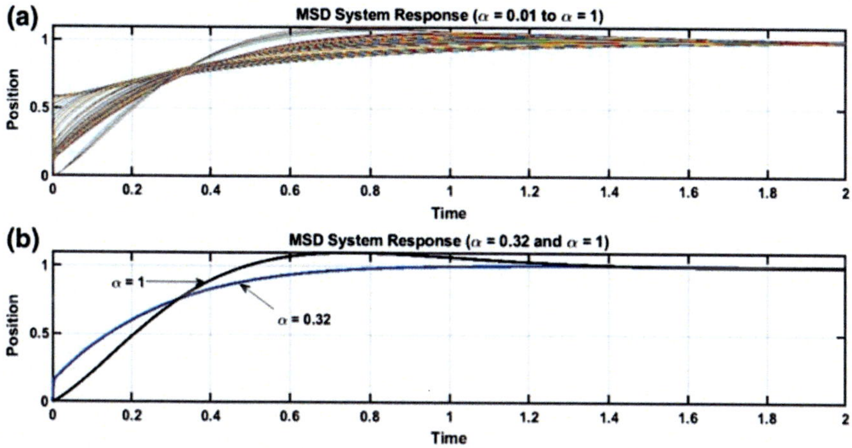

Fig. 6.4 PID Controller response of MSD system **a** for $\alpha = 0.01$ to $\alpha = 1$. **b** for $\alpha = 0.32$ (best fractional order system) and $\alpha = 1$ (integer order system)

Apply the proposed algorithm to control the velocity of a DC motor [33] whose transfer function is given as

$$TF(s) = \frac{K}{(R + Ls)(Js + b) + K^2}, \tag{6.11}$$

where, $J = 0.01$; $b = 0.1$; $K = 0.01$; $R = 1$; $L = 0.5$, and equivalent fractional order transfer function for fractional value α, is

$$TF(s) = \frac{0.01}{0.005s^{2\alpha} + 0.06s^{\alpha} + 0.1001}. \tag{6.12}$$

Stability behaviour of this fractional system is obtained by using the transformation (6.3). Considering $\alpha = 0.5$, the pole locations are at $-0.86 \pm 0.86j$ and $0.86 \pm 0.86j$ as shown in Fig. 6.5 which indicate that the system is stable.

For the instant, a step input is applied to the DC motor control system, and output is observed which is obtained by using the proposed algorithm. Let us consider this same system for two decimal place value of α. Then, the resultant PID controller responses are shown in Fig. 6.6.

Eliminating all the other PID controller responses from Fig. 6.6a which do not give desired response except for $\alpha = 1$ and $\alpha = 0.26$ (the best model) we get Fig. 6.6b. From Fig. 6.6b, the model corresponding to $\alpha = 0.26$ is the best fractional model which is obtained using the algorithm proposed. The settling time corresponding to this fractional model is 0.55 sec with 0.36% overshoot, which is less and better as compared to the settling time of the integer order model which is 1.3 sec and overshoot of 12.15%.

The fractional equivalent transfer function of this model found to be

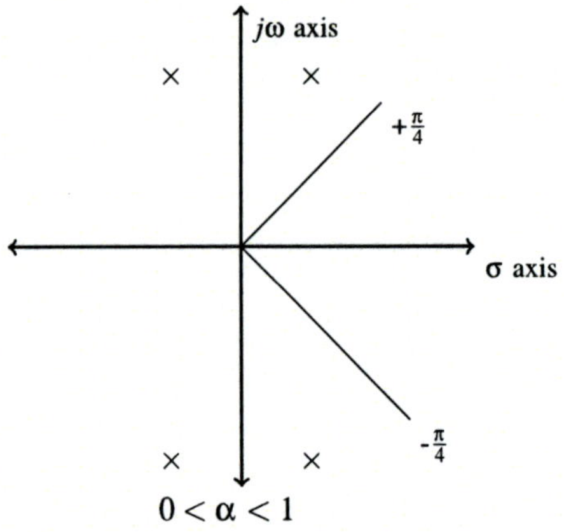

Fig. 6.5 Ω-domain analysis of the poles of a DC motor control system, all of which are in the left of the wedge shaped region, thereby ensuring stability

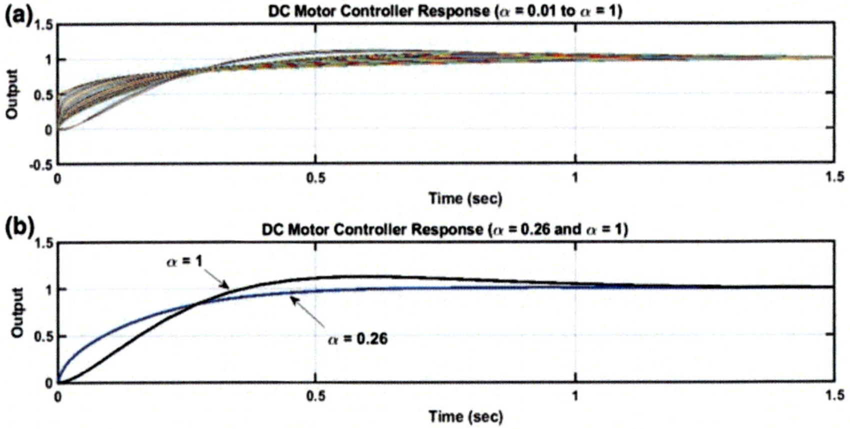

Fig. 6.6 PID Controller response of DC Motor control system **a** For $\alpha = 0.01$ to $\alpha = 1$. **b** For $\alpha = 0.26$ (best fractional model) and $\alpha = 1$ (integer order system)

$$TF(s) = \frac{0.01}{0.005s^{0.52} + 0.06s^{0.26} + 0.1001}, \tag{6.13}$$

and the approximated equivalent transfer function is shown as

$$G(s) = \frac{0.2638s + 0.002472}{s^2 + 3.353s + 0.02724}. \tag{6.14}$$

6.3 Simulation Results Analysis with Existing Fractional Models

Applying this algorithm to improve the fractional model of some pre-existing fractional order systems reported in [6, 27]. In [6], the authors reported a fractional order model of a system to design a PID control. In this section, the same system is considered and proposed algorithm is then applied to the system and found that the fractional order model detected by the proposed algorithm gives better control performance. The two improved fractional order model corresponding to the system reported in [6] when $\alpha = 0.1$ and $\alpha = 0.07$ respectively, are

$$G_1(s) = \frac{1}{0.7414s^{0.2} + 0.2313s^{0.1} + 1}, \quad G_2(s) = \frac{1}{0.7414s^{0.14} + 0.2313s^{0.07} + 1}. \tag{6.15}$$

Let us compare the performance results i.e. settling time and overshoot with the example of [6]. Figure 6.7 shows significant improvement in the response with fractional model and the proposed algorithm. The settling time of the example of [6] is 1 sec and overshoot is 20% while the settling time is 0.15 sec (for $\alpha = 0.1$) and 0.1314 sec (for $\alpha = 0.07$) with no overshoot in the improved fractional model obtained by proposed algorithm. There is an improvement in the performance by 85% when $\alpha = 0.1$ and 86.86% when $\alpha = 0.07$ (Refer Table 6.1).

Taking the example of a heating furnace model [27] with fractional behavior and its corresponding fractional model is shown as

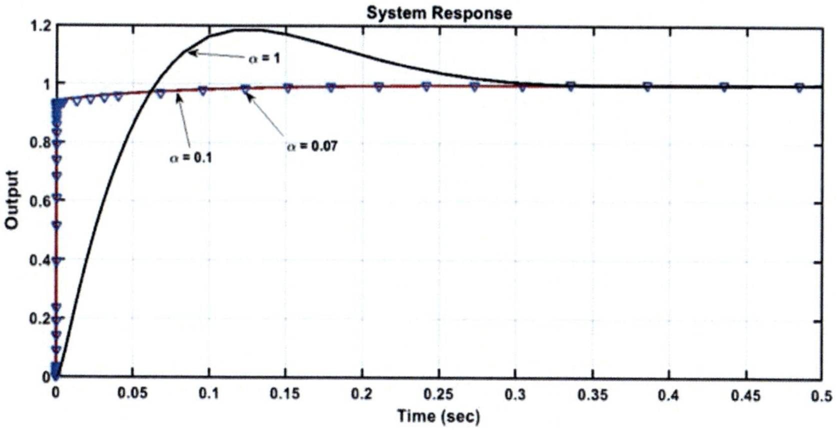

Fig. 6.7 Response of the improved fractional order model of paper [6], example 2, corresponding to $\alpha = 0.1$, $\alpha = 0.07$, in comparison to integer order model of $\alpha = 1$

Table 6.1 Comparison of the specifications with Chunna Zhao et al. model [6]

Specifications	Chunna Zhao's model	Proposed algorithm	
		$\alpha = 0.1$	$\alpha = 0.07$
Settling time (sec)	1	0.15 (improved by 85%)	0.1314 (improved by 86.86%)
% Overshoot	20	No	No

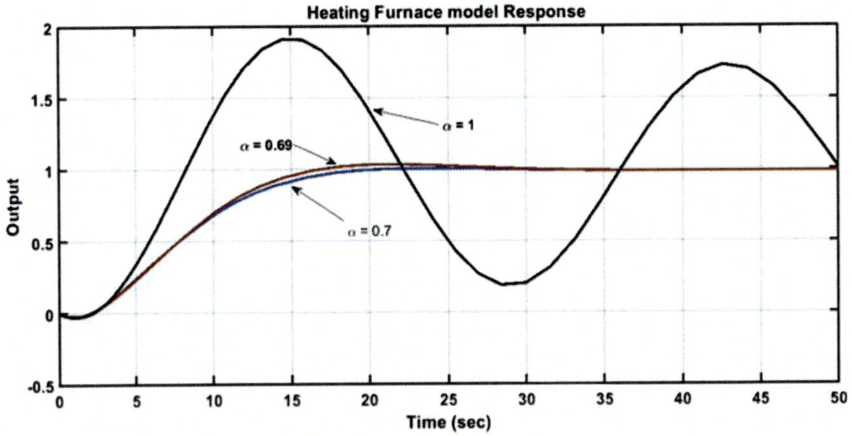

Fig. 6.8 Responses of the improved fractional order model of heating furnace system with the proposed algorithm and the best fractional order models corresponding to $\alpha = 0.7$, and $\alpha = 0.69$, as compared to integer order model ($\alpha = 1$)

$$G(s) = \frac{1}{14994s^{1.31} + 6009.5s^{0.97} + 1.69}. \tag{6.16}$$

When the proposed algorithm is applied to the system defined by (6.16) then two new improved fractional model of this system are obtained as

$$G_1(s) = \frac{1}{73043s^{1.4} + 4893s^{0.7} + 1.93}, \tag{6.17}$$

$$G_2(s) = \frac{1}{73043s^{1.38} + 4893s^{0.69} + 1.93}. \tag{6.18}$$

The response corresponding to these two systems are very much improved in terms of settling time and overshoot. The settling time of the example of [27] is 400 sec and overshoot is 18% while the settling time is 24.9 sec (for $\alpha = 0.7$) and 14 sec (for $\alpha = 0.69$) with negligible overshoot in the improved fractional model obtained by proposed algorithm as shown in Fig. 6.8.

There is significant improvement in the performance by 93.8% when $\alpha = 0.7$ and 96.5% when $\alpha = 0.69$ of the heating furnace. We depicts that the % improvement

Table 6.2 Comparison of the specifications with Mingda's model [27]

Specifications	Mingda's model		Proposed algorithm	
	ADRC control	PID control	$\alpha = 0.7$	$\alpha = 0.69$
Settling time (sec)	50	400	24.9 (improved by 50.2% wrt ADRC and 93.8% wrt PID)	14 (improved by 72% wrt ADRC and 96.5% wrt PID)
% Overshoot	No	18	3.12	0.3

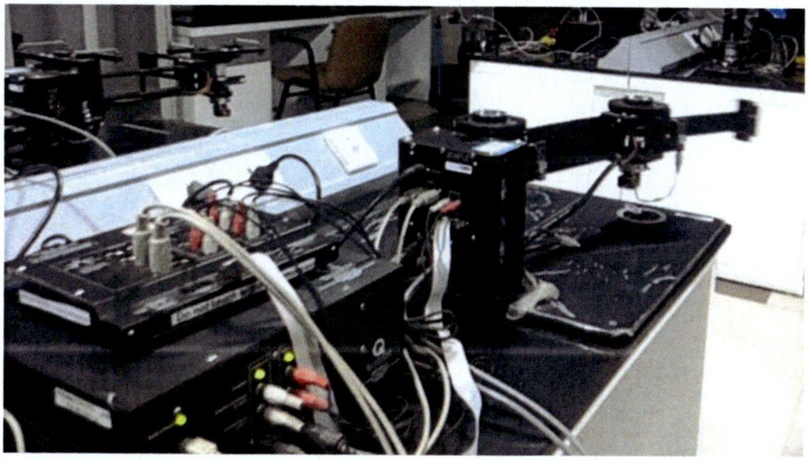

Fig. 6.9 Experimental setup of 2DOF serial flexible link robotic manipulator

in settling time of (6.17) and (6.18) is much improved, by more than twofold (Refer Table 6.2). If we consider fractional model corresponding to two decimal places (see (6.19)) then this model will be perfect to design controller.

6.4 Experimental Results Analysis on Robotic Manipulators

In this section, serial link and serial joint robotic manipulators of two degrees of freedom, are examined, and the experimental setups are shown in Figs. 6.9 and 6.10 respectively.

The modeling equations are obtained from the user manual provided by Quanser Inc. [30, 31]. Other works using similar Quanser setups are available in literature [29]. The matrices in state-space form for first stage of two degrees of freedom flexible link Robotic manipulator are

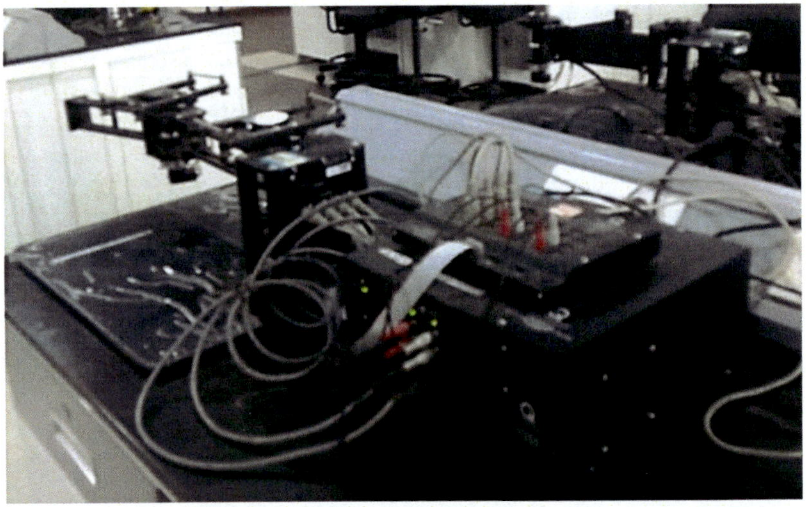

Fig. 6.10 Experimental setup of 2DOF serial flexible joint robotic manipulator

$$A_{1l} = \begin{pmatrix} 0 & 0 & 1 & 0 \\ 0 & 0 & 0 & 1 \\ 0 & 628.88 & -62.95 & 0 \\ 0 & -863.33 & 62.95 & 0 \end{pmatrix}, \; B_{1l} = \begin{pmatrix} 0 \\ 0 \\ 140.47 \\ -140.47 \end{pmatrix}, \tag{6.19}$$

and the state space matrices for the second stage of this manipulator are

$$A_{2l} = \begin{pmatrix} 0 & 0 & 1 & 0 \\ 0 & 0 & 0 & 1 \\ 0 & 2271.1 & -496.8 & 28.4 \\ 0 & -3336.2 & 496.8 & -41.6 \end{pmatrix}, \; B_{2l} = \begin{pmatrix} 0 \\ 0 \\ 288.1224 \\ -288.1224 \end{pmatrix}. \tag{6.20}$$

The matrices for first stage of two degrees of freedom flexible joint Robotic manipulator are

$$A_{1j} = \begin{pmatrix} 0 & 0 & 1 & 0 \\ 0 & 0 & 0 & 1 \\ -141.2 & 141.2 & -70.6 & 0 \\ 39 & -39 & 0 & -0.3 \end{pmatrix}, \; B_{1j} = \begin{pmatrix} 0 \\ 0 \\ 140 \\ 0 \end{pmatrix}, \tag{6.21}$$

and the state space matrices for the second stage of this manipulator are

$$A_{2j} = \begin{pmatrix} 0 & 0 & 1 & 0 \\ 0 & 0 & 0 & 1 \\ -1141 & 1141 & -142.6 & 0 \\ 373 & -373 & 0 & -2.6 \end{pmatrix}, \; B_{2j} = \begin{pmatrix} 0 \\ 0 \\ 248.2 \\ 0 \end{pmatrix}. \tag{6.22}$$

The equivalent integer order transfer functions obtained from the state-space matrices (6.20)–(6.23) are

$$\theta_{11l}(s) = \frac{140.5s^2 + 3.293 \times 10^4}{s^4 + 62.95s^3 + 863.3s^2 + 1.476 \times 10^4 s}, \tag{6.23}$$

$$\theta_{21l}(s) = \frac{290s^2 + 3800s + 306880}{s^4 + 540s^3 + 9891s^2 + 52914s}, \tag{6.24}$$

where $\theta_{11l}(s)$ is the angular position of the first link and $\theta_{21l}(s)$ is the angular position of the second link, and similarly the transfer function of this manipulator can be obtained as

$$\theta_{11j}(s) = \frac{140s^2 + 42.77s + 5470}{s^4 + 70.91s^3 + 201.8s^2 + 2801s}, \tag{6.25}$$

$$\theta_{21j}(s) = \frac{248.2s^2 + 652.8s + 92560}{s^4 + 145.2s^3 + 1889s^2 + 56200s}, \tag{6.26}$$

where $\theta_{11j}(s)$ is the angular position of the first joint and $\theta_{21j}(s)$ is the angular position of the second joint. Let us obtain the fractional equivalent model corresponding to these robotic manipulators. Taking (6.23) and (6.24), the generalized fractional model can be obtained as

$$\theta_{11l}(s^\alpha) = \frac{140.5s^{2\alpha} + 3.293 \times 10^4}{s^{4\alpha} + 62.95s^{3\alpha} + 863.3s^{2\alpha} + 1.476 \times 10^4 s^\alpha}, \tag{6.27}$$

$$\theta_{21l}(s^\alpha) = \frac{290s^{2\alpha} + 3800s + 306880}{s^{4\alpha} + 540s^{3\alpha} + 9891s^{2\alpha} + 52914s^\alpha}. \tag{6.28}$$

Applying the proposed algorithm, we get an accurate fractional model. The value of α obtained for Link 1 is $\alpha = 0.3$ and for Link 2, it is $\alpha = 0.9$. Hence, the approximated fractional model of these systems are

$$\bar{\theta}_{11l}(s^{0.3}) = \frac{3.74 \times 10^{13}}{s^2 + 8.54 \times 10^{12}s + 1.57 \times 10^{10}}, \tag{6.29}$$

$$\bar{\theta}_{21l}(s^{0.9}) = \frac{2.6 \times 10^{13}}{s^2 + 7.64 \times 10^{12}s + 8.97 \times 10^9}. \tag{6.30}$$

Similarly, the fractional model of the flexible joint robotic manipulator is obtained when α value of Joint 1 is $\alpha = 0.9$ and α value of Joint 2 is $\alpha = 0.9$, and the equivalent approximated fractional model is expressed as

$$\bar{\theta}_{11j}(s^{0.9}) = \frac{1.62 \times 10^{13}}{s^2 + 1.41 \times 10^{13}s + 1.7 \times 10^{10}}, \tag{6.31}$$

$$+\bar{\theta}_{21j}(s^{0.9}) = \frac{1.4 \times 10^{13}}{s^2 + 1.45 \times 10^{13}s + 1.7 \times 10^{10}}. \tag{6.32}$$

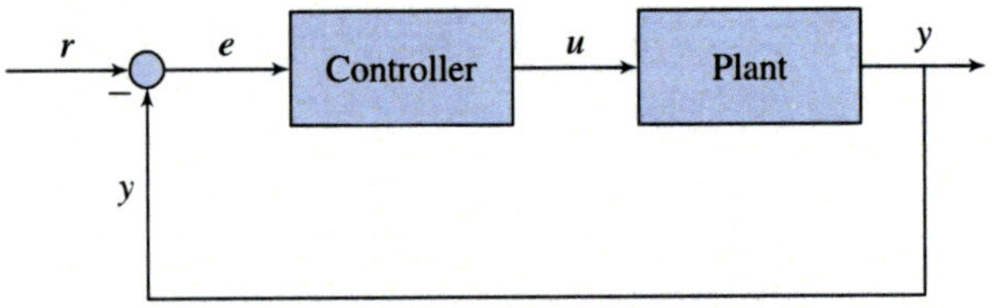

Fig. 6.11 Controller design strategy for a plant

6.4.1 *Fractional Controller Design for Integer Order Models*

The schematic of controller design is shown in Fig. 6.11 for the plant represented by (6.23)–(6.26).

An appropriate fractional order controller is represented as

$$C(s) = K_p + K_i s^{-\alpha} + K_d s^{\beta}, \tag{6.33}$$

with 5 tunable variables: K_p (proportional gain), K_i (integral gain), K_d (differentiator gain), α (integrator exponent) and β (differentiator exponent).

As shown in Fig. 6.9, the two links are not coupled and hence the controller design strategy for the angular position of the two links are independent. The aim is to control the angular position of Link 1 and Link 2 at desired positions. The desired angular position of Link 1 is to track a square wave which varies at an angle of ±15° periodically after 5 sec, and the desired angular position of Link 2 is to track a square wave which varies at an angle of ±10° periodically after 5 sec. A delay of 3 sec is provided in between the tracking of the desired positions of Link 1 and Link 2. Four cases are considered to design the control strategy depending on what angular position is to be controlled:

1. Link 1 Fractional Controller and Link 2 Fractional Controller
2. Link 1 Integer Controller and Link 2 Fractional Controller
3. Link 1 Fractional Controller and Link 2 Integer Controller
4. Link 1 Integer Controller and Link 2 Integer Controller.

Let us consider Case 1 where the angular position of Link 1 is controlled by a fractional controller given as

$$C(s)_{Link1} = 4.5 + 10s^{-0.1} + 0.23s^{0.5}, \tag{6.34}$$

and the angular position of Link 2 is controlled by a fractional controller given as

$$C(s)_{Link2} = 10.6 + 8s^{-0.5} - 0.25s^{0.5}. \tag{6.35}$$

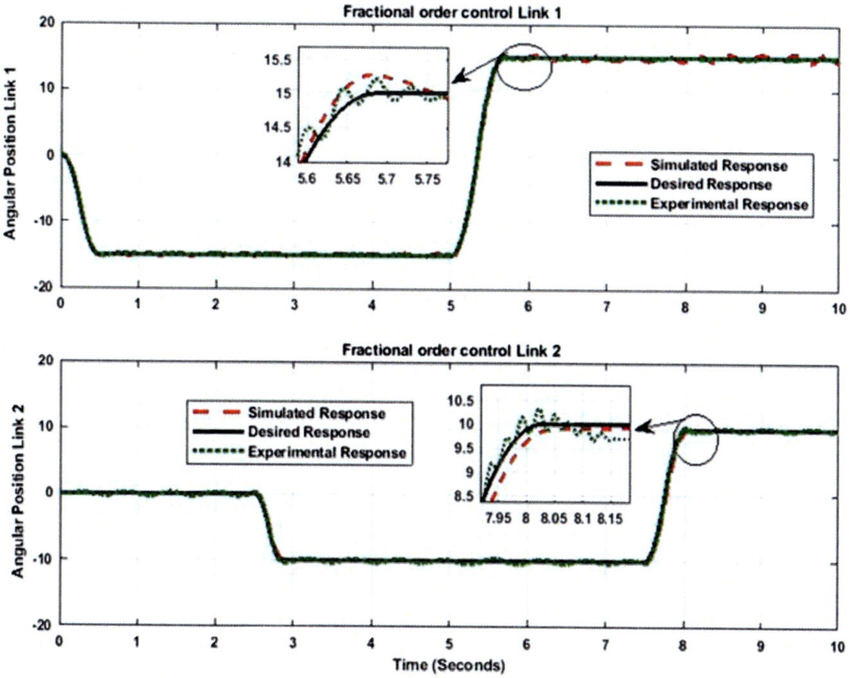

Fig. 6.12 The simulated and experimental fractional order control applied to integer order models of two degree of freedom serial flexible link robotic manipulator

The simulated and experimental responses are shown in Fig. 6.12. As clearly seen, the responses of both the links track the desired response with negligible overshoot. In this way, let us compare the results obtained for these cases of controller design to check the effectiveness of the controller.

From the responses of the angular position control of Link 1 and angular position control of Link 2 shown in Fig. 6.13 we can deduce that the fractional order controller performs better (in terms of settling time and overshoot) when compared to the integer order controller. As shown in Fig. 6.10, the two Joints are not coupled and hence the controller design strategy for the angular position of the two Joints are independent. Four cases have been considered to design the control strategy depending on what angular position is to be controlled:

1. Joint 1 Fractional Controller and Joint 2 Fractional Controller
2. Joint 1 Integer Controller and Joint 2 Fractional Controller
3. Joint 1 Fractional Controller and Joint 2 Integer Controller
4. Joint 1 Integer Controller and Joint 2 Integer Controller.

Let us consider Case 1 where the angular position of the Joint 1 is controlled by a fractional controller of the form

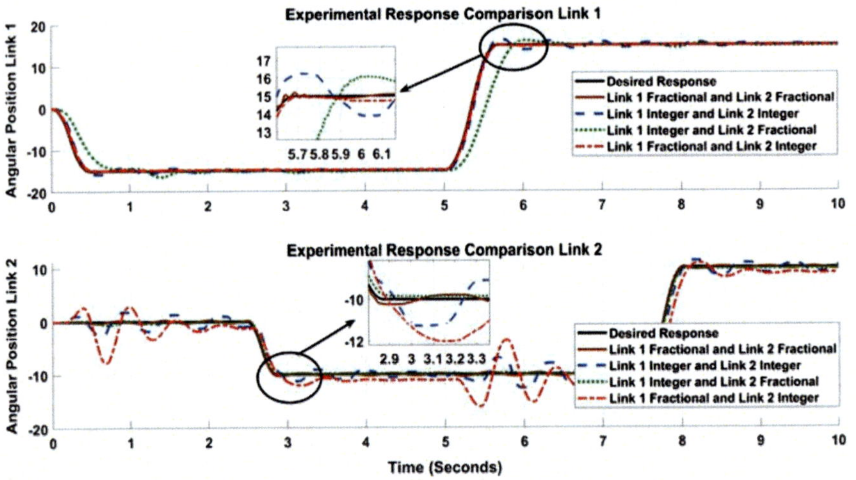

Fig. 6.13 Experimental controlled response comparison shows that the fractional controlled responses for both links is the best as compared to fractional order controller for one of the links and integer controller for the other, or integer controller for both the links

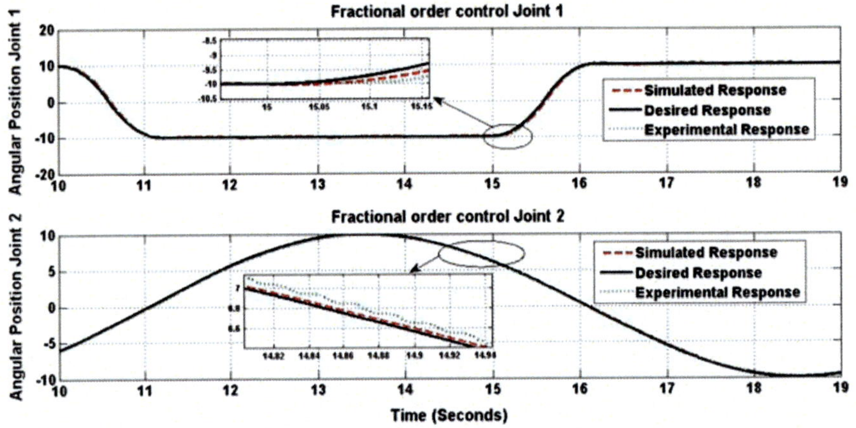

Fig. 6.14 The simulated and experimental fractional order control applied to integer order models of two degree of freedom serial flexible joint robotic manipulator

$$C(s)_{Joint1} = 20 + 2s^{-0.5} + 0.1s^{0.5}, \tag{6.36}$$

and the angular position of the Joint 2 is controlled by a fractional controller given as

$$C(s)_{Joint2} = 16 + 37s^{-0.5} + 0.3s^{0.5}. \tag{6.37}$$

The simulated and experimental responses are shown in Fig. 6.14. As clearly seen, the responses of both the joints track the desired response with negligible overshoot.

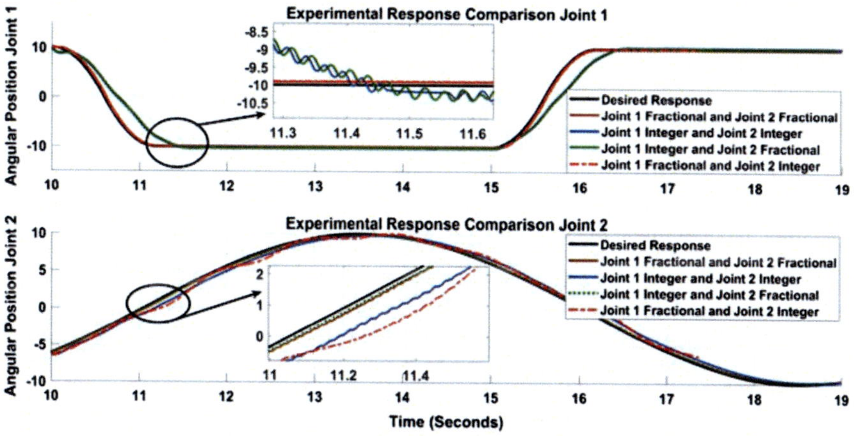

Fig. 6.15 Experimental results show that the fractional controlled responses for both joints, are the best as compared to fractional order controller for one of the joints and integer controller for the other, or integer controller for both the joints

From the responses of the angular position control of joint 1 and angular position control of joint 2 shown in Fig. 6.15, we can deduce that the fractional order controller performs better (in terms of settling time and overshoot) when compared to the integer order controller.

6.4.2 Integer Controller for Fractional Order Models

In this section, an integer order controller is designed for fractional order model corresponding to two degrees of freedom serial flexible fink and joint robotic manipulators. There are no fractional order models present in literature corresponding to such robotic manipulators. Hence, by using the proposed algorithm, fractional models of the manipulators are obtained, and then the appropriate integer order controller is designed by tuning the PID gains accordingly for desired response. For experimental validation, the tuned gain values of the controller obtained from the simulation is applied to the experimental setup of the robotic manipulators. The desired and experimental response of these robotic manipulator systems are shown in Fig. 6.16. The tuned controller of Link 1 takes the form

$$C(s)_{Link1} = 8.8 + \frac{20}{s} - 0.01s. \tag{6.38}$$

The tuned controller of Link 2 takes the form

$$C(s)_{Link2} = 11 + \frac{3}{s} - 0.06s. \tag{6.39}$$

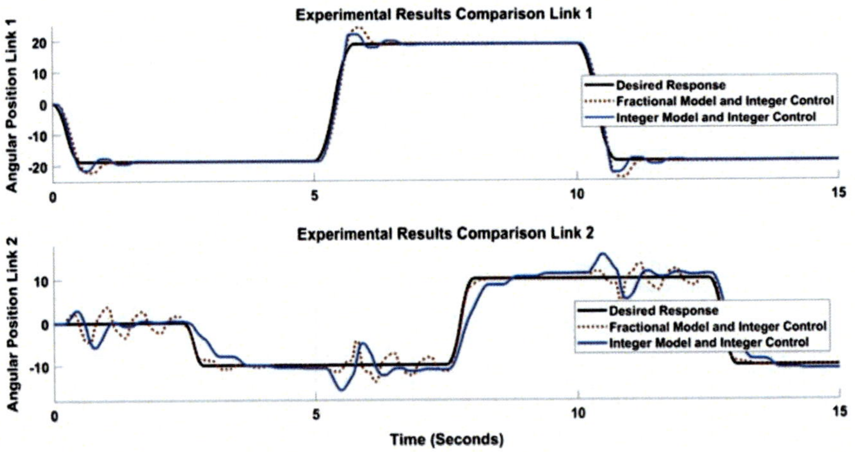

Fig. 6.16 Experimental controlled response comparison shows that the integer order controller designed for fractional order model of both the links is best as compared to the integer order controller designed for integer order model of the links

It can be seen that the controller design for integer model takes about 2.2 sec to track the desired response. The desired and experimental response of the flexible joint robotic manipulator system is shown in Fig. 6.17. The tuned controller of Joint 1 takes the form

$$C(s)_{Joint1} = 21 + \frac{95}{s} - 0.2s. \tag{6.40}$$

The tuned controller of Joint 2 takes the form

$$C(s)_{Joint2} = 25 + \frac{113}{s} - 0.22s. \tag{6.41}$$

It can be seen that the controller design for integer model provides a response with non-zero steady state error.

6.4.3 *Fractional Controller Design for Fractional Order Model*

Comparing the results obtained in Fig. 6.18, it is observed that the Fractional order controller designed for the fractional order model of Link 1 tracks the desired response in minimum time while the integer order controller faces certain steady state error issues [32]. For Link 2, the fractional order controller for fractional modal effectively tracks the desired response while the integer order controller response oscillates along the desired response.

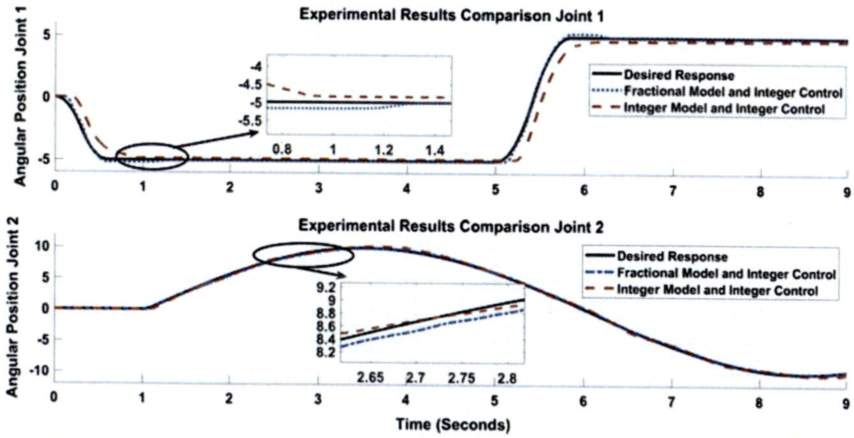

Fig. 6.17 Experimental controlled response comparison shows that the integer order controller designed for fractional order model of both the joints is best as compared to the integer order controller designed for integer order model of the joints

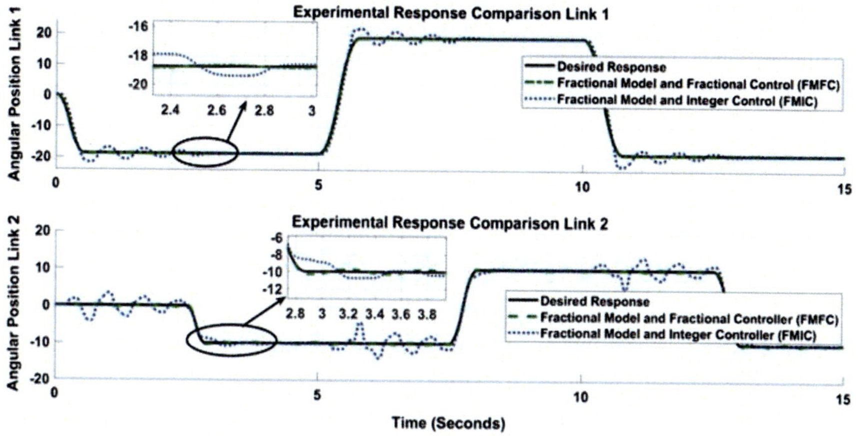

Fig. 6.18 Experimental controlled response comparison shows that the fractional order controller designed for fractional order model of both the links is best as compared to the integer order controller designed for fractional order model of the links

The fractional order controller tuned for Link 1 takes the form

$$C(s)_{Link1} = 70 + 50s^{-0.2} + 0.23s^{0.8}, \tag{6.42}$$

and the fractional order controller tuned for Link 2 takes the form

$$C(s)_{Link2} = 15.5 + 90s^{-0.1} + 0.25s^{0.9}. \tag{6.43}$$

Similarly, a fractional order controller can be designed for fractional order model of the two degree of freedom flexible joint robotic manipulator.

This chapter proposes an algorithm to detect the best suitable fractional model to design a controller that can track the desired response. In this work, different second order systems are considered wherein the proposed algorithm is applied, and suitable fractional order models are reported. Furthermore, the proposed algorithm is applied to two pre-existing fractional order systems and the algorithm suggested that the improved fractional model corresponding to these systems improve the transient responses by twofold. The proposed algorithm may be useful in many industrial applications where excellent control of stability is required, e.g., in controlling robotic manipulators, controlling of under-actuated and fully actuated mechanical systems. This algorithm can also be used in chemical processes to detect the best model for a chemical reaction. In the next chapter, the experimental validation of the results obtained until this chapter is presented.

This chapter also proposed a new fractional model of two degrees of freedom serial link and serial joint robotic manipulators. It is observed that a fractional order controller for a fractional order robotic manipulator gives the best response as per experimental validation, compared to the cases wherein fractional order controller is applied to an integer order model or an integer order controller is applied to a fractional order model. The examined results show significantly improved performance using the fractional controller and fractional model and this shows the strong evidence of reliability of the fractional model in control system design.

References

1. Deb, D., Tao, G., Burkholder, J.O., Smith, D.R.: Adaptive compensation control of synthetic jet actuator arrays for airfoil virtual shaping. J. Aircr. **44**(2), 616–626 (2007)
2. Monje, C.A., Chen, Y., Vinagre, B.M., Xue, D., Feliu-Batlle, V.: Fractional-Order Systems and Controls: Fundamentals and Applications. Springer Science & Business Media, Berlin (2010)
3. Machado, J.T., Lopes, A.M.: Ranking the scientific output of researchers in fractional calculus. Fract. Calc. Appl. Anal. **22**(1), 11–26 (2019)
4. Monje, C.A., Calderon, A.J., Vinagre, B.M., Chen, Y., Feliu, V.: On fractional PI controllers: some tuning rules for robustness to plant uncertainties. Nonlinear Dyn. **38**(1–4), 369–381 (2004)
5. Monje, C.A., Vinagre, B.M., Chen, Y.Q., Feliu, V., Lanusse, P., Sabatier, J.: Proposals for fractional PI D tuning. In: The First IFAC Symposium on Fractional Differentiation and its Applications, vol. 38, pp. 369–381(2004)
6. Zhao, C., Xue, D., Chen, Y.: A fractional order PID tuning algorithm for a class of fractional order plants. In: IEEE International Conference Mechatronics and Automation, vol. 1. IEEE, pp. 216–221 (2005)
7. Monje, C.A., Vinagre, B.M., Feliu, V., Chen, Y.: Tuning and auto-tuning of fractional order controllers for industry applications. Control Eng. Pract. **16**(7), 798–812 (2008)
8. Li, H., Luo, Y., Chen, Y.: A fractional order proportional and derivative (FOPD) motion controller: tuning rule and experiments. IEEE Trans. Control Syst. Technol. **18**(2), 516–520 (2009)
9. Senberber, H., Bagis, A.: Fractional PID controller design for fractional order systems using ABC algorithm. In: Electronics. IEEE, pp. 1–7 (2017)

10. Padula, F., Visioli, A.: Tuning rules for optimal PID and fractional-order PID controllers. J. Process Control **21**(1), 69–81 (2011)
11. Zhang, B., Pi, Y., Luo, Y.: Fractional order sliding-mode control based on parameters auto-tuning for velocity control of permanent magnet synchronous motor. ISA Trans. **51**(5), 649–656 (2012)
12. Badri, V., Tavazoei, M.S.: On tuning fractional order [proportional-derivative] controllers for a class of fractional order systems. Automatica **49**(7), 2297–2301 (2013)
13. Senberber, H., Bagis, A.: Fractional PID controller design for fractional order systems using ABC algorithm. In: 2017 Electronics. IEEE, pp. 1–7 (2017)
14. Chen, Y., Bhaskaran, T., Xue, D.: Practical tuning rule development for fractional order proportional and integral controllers. J. Comput. Nonlinear Dyn. **3**(2), 021403 (2008)
15. Hamamci, S.E.: An algorithm for stabilization of fractional-order time delay systems using fractional-order PID controllers. IEEE Trans. Autom. Control **52**(10), 1964–1969 (2007)
16. Singh, A., Agrawal, H.: A fractional model predictive control design for 2-d gantry crane system. J. Eng. Sci. Technol. **13**(7), 2224–2235 (2018)
17. Srivastava, T., Singh, A.P., Agarwal, H.: Modeling the under-actuated mechanical system with fractional order derivative. Progr. Fract. Differ. Appl. **1**(1), 57–64 (2015)
18. Singh, A.P., Kazi, F.S., Singh, N.M., Srivastava, P.: PI D controller design for underactuated mechanical systems. In: 2012 12th International Conference on Control Automation Robotics & Vision (ICARCV). IEEE, pp. 1654–1658 (2012)
19. AboBakr, A., Said, L.A., Madian, A.H., Elwakil, A.S., Radwan, A.G.: Experimental comparison of integer/fractional-order electrical models of plant. AEU-Int. J. Electron. Commun. **80**, 1–9 (2017)
20. Beltempo, A., Zingales, M., Bursi, O.S., Deseri, L.: A fractional-order model for aging materials: an application to concrete. Int. J. Solids Struct. **138**, 13–23 (2018)
21. Martynyuk, V., Ortigueira, M., Fedula, M., Savenko, O.: Methodology of electrochemical capacitor quality control with fractional order model. AEU-Int. J. Electron. Commun. **91**, 118–124 (2018)
22. Chen, Y., Petras, I., Xue, D.: Fractional order control-a tutorial. In: 2009 American control conference. IEEE, pp. 1397–1411 (2009)
23. Shalaby, R., El-Hossainy, M., Abo-Zalam, B.: Fractional order modeling and control for under-actuated inverted pendulum. Commun. Nonlinear Sci. Numer. Simul. **74**, 97–121 (2019)
24. Ortigueira, M.D., Valério, D., Machado, J.T.: Variable order fractional systems. Commun. Nonlinear Sci. Numer. Simul. **71**, 231–243 (2019)
25. Machado, J.T., Lopes, A.M.: Fractional-order modeling of a diode. Commun. Nonlinear Sci. Numer. Simul. **70**, 343–353 (2019)
26. Teodoro, G.S., Machado, J.T., De Oliveira, E.C.: A review of definitions of fractional derivatives and other operators. J. Comput. Phys. **388**, 195–208 (2019)
27. Li, M., Li, D., Wang, J., Zhao, C.: Active disturbance rejection control for fractional-order system. ISA Trans. **52**, 365–374 (2013)
28. Bandyopadhyay, B., Kamal, S.: Solution, stability and realization of fractional order differential equation. In: Stabilization and Control of Fractional Order Systems: A Sliding Mode Approach. Springer, Cham, pp. 55–90 (2015)
29. Patel, R., Deb, D., Modi, H., Shah, S.: Adaptive backstepping control scheme with integral action for quanser 2-dof helicopter. In: 2017 International Conference on Advances in Computing, Communications and Informatics (ICACCI). IEEE, pp. 571–577 (2017)
30. Quanser, Dynamic equations for the first stage of the serial flexible link340(2dsfl) robot and serial flexible joint (2dsfj) robot. In: Maple Worksheet341or HTML File.342 [34]
31. Quanser, Dynamic equations for the second stage of the serial flexi-343ble link (2dsfl) robot and serial flexible joint (2dsfj) robot, in: Maple344Worksheet or HTML File

32. Singh, A., Deb, D., Agarwal, H.: On selection of improved fractional model and control of different systems with experimental validation. Commun. Nonlinear Sci. Numer. Simul. **79**, 104902 (2019). https://doi.org/10.1016/j.cnsns.2019.104902
33. Deb, D.: Nonlinear adaptive control of vehicular radar servo systems with system and sensor uncertainties. In: 2011 IEEE Applied Electromagnetics Conference (AEMC) (2011). https://doi.org/10.1109/aemc.2011.6256910

Chapter 7
Model Reference Adaptive Fractional Order Controller Design

Various controller design strategies have been studied in the past for under-actuated and fully actuated robotic systems to control a specific task in the best possible manner. This chapter proposes an adaptive fractional order (FO) controller design for under-actuated and fully actuated robotic manipulators. Adaptive controllers can adapt to a particular uncertainty inside the plant and give the best possible controlled response [1]. The change may due to variation of an individual parameter or may be due to disturbances. A suitable controller is supposed to adapt to these uncertainties and give the best results.

7.1 Introduction

Adaptive controllers have been used in diverse applications such as systems biology [2], aircraft flight control [3–7], batteries [8], vehicular traffic control [9, 10], wireless networked systems [11], and renewable energy systems like solar panels [12, 13], wind turbines [14] and microbial fuel cells [15–17], and the list is endless. A few related papers and their contributions in the context of adaptive control, are highlighted herewith. Nath et al. have designed robust observer-based adaptive controller for an intravenous glucose tolerance nonlinear model of Type 1 Diabetes Mellitus (T1DM) patients wherein an adaptive control technique uses the Lyapunov's second method for desired closed-loop performance of the uncertain system [18]. The simulation results depict that the design can avoid hypoglycemia (glucose concentration level below 50 mg/dl) and also brings down the glucose concentration safely within two hours [19]. Advanced wind measuring systems like Light Detection and Ranging is useful for wind farm wake management using adaptive control of wake center estimation and to steer the wake center to the desired value, for wake control

A. P. Singh et al., *Fractional Modeling and Controller Design of Robotic Manipulators*, Intelligent Systems Reference Library 194,
https://doi.org/10.1007/978-3-030-58247-0_7

simulations. The estimated effective wind speed is used to model wind farms in the form of transfer functions [20].

Next-generation airplanes with an array of active flow control devices called synthetic jet actuators [21, 22] are deployed for improved aerodynamic performance, but due to the inherent nonlinearities [23], there is a need for an adaptive compensation scheme at low angles of attack [24]. A set of adaptive inverse models of the actuators is employed for cancelling the effect of the jet nonlinearities. A nonlinear state feedback control law is designed, wherein a set of intermediate states is used as controls for other aircraft states. Parameter projection-based adaptive laws ensure closed-loop stability [25]. Kapoor et al. present an adaptive control mechanism for failure compensation for a coaxial rotor helicopter while addressing the problem of instability due to rotor damage in the form of degradation or unexpected trust variations from the propellers [6]. The problem of instability along the yaw and the altitude axis for coaxial rotor helicopters under propeller failure is modeled, and a suitable adaptive control scheme is proposed to stabilize the helicopter's states under propeller failure [26].

Patel and Deb propose a novel adaptive backstepping control strategy for a single chamber microbial fuel cell [27] with pure culture, wherein an uncertain parameter (microorganisms growth rate) is estimated online through the adaptive control law and the influent substrate flow rate is considered as manipulated input variable [28, 29]. Later, Patel and Deb also developed an adaptive control technique for single chamber MFC with two microorganisms. In this control technique, the initial substrate concentration is considered as a manipulated input variable and two uncertain parameters are estimated [30].

Wei has applied adaptive control design for robotic manipulators based on Lyapunov stability theory and hyperstability theory. For the interaction of the robot with the environment, force control is considered to ensure a safe operation [31]. He et al. Propose an adaptive impedance control is developed for a n-link robotic manipulator with input saturation which is handled through an auxiliary system, and uncertainties by employing radial basis function-based adaptive controller [32]. An adaptive second-order sliding mode control design for a two-link robotic manipulator with an effective adaptive tuning law is presented even without prior knowledge of uncertainty bounds of a robotic manipulator and the control signal is chattering free and requires less control effort [33]. This work is then extended to remain chattering-free under inertia uncertainties and external disturbance, and achieve fast convergence [34]. Finite-time stabilization and control of a n-degree of freedom (DOF) robotic manipulator are looked at while considering uncertainties, and external disturbances by integrating adaptive backstepping control approach into fractional-order controller design for an integer-order dynamic system [35]. Singh et al. have already validated experimentally an improved FO model of robotic manipulators [36] and now this technique needs to be extended to adaptive FO systems.

Next, a few of such adaptive control methodologies in fractional order are applied to different under-actuated and fully actuated robotic systems to account for changes in parametric uncertainties. The considered under-actuated systems are (i) pendulum-cart system, (ii) 2D gantry crane system, (iii) 2DOF serial link robotic manipulator, and the fully actuated systems are (i) single rigid link robotic manipulator, (ii) missile launching vehicle (MLV) system. The adaptation parameters are found by minimizing a cost function in each case. Numerical simulations show the superiority of adaptive fractional order controller for under-actuated and fully-actuated robotic systems.

7.2 Adaptive FO Controller Design for Pendulum on a Cart System

Let us consider the same model which has been considered in Chap. 2 and design an adaptive fractional PID controller based on model reference. POAC system is an under-actuated system because this system has fewer actuators when compared with their degrees of freedom. Recalling the state-space model of POAC

$$\frac{d}{dt}(\delta \underline{z}) = \begin{pmatrix} 0 & 1 & 0 & 0 \\ 0 & \frac{-B}{M} & \frac{-mg}{M} & 0 \\ 0 & 0 & 0 & 1 \\ 0 & \frac{B}{Ml} & \frac{(M+m)g}{Ml} & 0 \end{pmatrix} \delta \underline{z} + \begin{pmatrix} 0 \\ \frac{1}{M} \\ 0 \\ \frac{-1}{Ml} \end{pmatrix} \delta u \tag{7.1}$$

with output equation as

$$y = \begin{pmatrix} 1 & 0 & 0 & 0 \\ 0 & 0 & 1 & 0 \end{pmatrix} \begin{pmatrix} x \\ \dot{x} \\ \theta \\ \dot{\theta} \end{pmatrix}. \tag{7.2}$$

Assuming, there is no friction $B = 0$ and $m = 0.2\,\text{Kg}$, $M = 0.5\,\text{Kg}$, $l = 1\,\text{m}$ and $g = 10\,\text{m/s}^2$. Then from (7.1) and (7.2), we get

$$A = \begin{pmatrix} 0 & 1 & 0 & 0 \\ 0 & 0 & -4 & 0 \\ 0 & 0 & 0 & 1 \\ 0 & 0 & 14 & 0 \end{pmatrix}, \quad B = \begin{pmatrix} 0 \\ 2 \\ 0 \\ -2 \end{pmatrix}, \quad C = \begin{pmatrix} 1 & 0 & 0 & 0 \\ 0 & 0 & 1 & 0 \end{pmatrix}. \tag{7.3}$$

From (7.3), it is easy to follow

$$X(s) = \frac{2s^2 - 36}{s^4 - 14s^2}, \quad \theta(s) = \frac{2s^2}{s^4 - 14s^2}. \tag{7.4}$$

This system is under-actuated having only one plant actuation and two degrees of freedom i.e., position and angle. Let us design an adaptive mechanism which adapts as per the desired position. The adaptive controller block diagram is shown in Fig. 7.1.

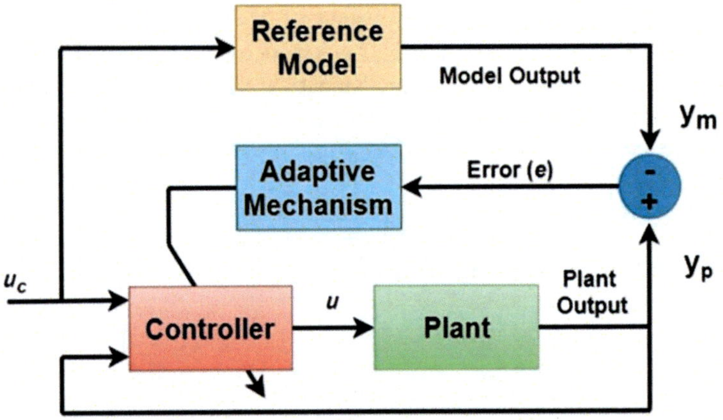

Fig. 7.1 Model reference-adaptive controller block diagram

The most popular approach to design adaptive controller is Model Reference Adaptive Control (MRAC), which track the desired performance by the choice of a reference model. It is required to adjust the parameters and hence the adjustment of parameters are achieved utilizing the error between the output of the reference model and plant output [37]. The reference model is chosen such that it generates the desired response which is to be tracked by the plant output. The change in the desired response to the plant output can be obtained from Fig. 7.1 as

$$e = y_p - y_m, \tag{7.5}$$

where plant output is y_p and model output is y_m. If there is a difference in the output responses of the reference model and output response of plan, then it alters the amplitude of e. The MRAC changes the adaptation parameter $x(t)$ by minimizing a cost function by using an adaptation rule [38]:

$$J(x) = \frac{1}{2}e^2. \tag{7.6}$$

The adaptation rule of the MRAC for the change in the adaptation parameter towards the direction of negative gradient of $J(x)$ [39],

$$\frac{dx}{dt} = -\gamma\frac{dJ}{dx} = -\gamma e\frac{de}{dx}, \tag{7.7}$$

where γ is adaptation gain. Recall (7.4), and from Fig. 7.1, it is easy to follow

$$u = x * u_c, \tag{7.8}$$

where x is adaptation parameter and from (7.6),

$$E(s) = G_p(s)U(s) - G_m(s)U_c(s), \tag{7.9}$$

where $G_p(s)$ is the plant model and $G_m(s)$ is reference model. Then, the output signals and error signals are

$$y_p = G_pU = \left(\frac{2s^2 - 36}{s^4 - 14s^2}\right) xu_c, \;\; y_m = G_m u, \;\; e = \left(\frac{2s^2 - 36}{s^4 - 14s^2}\right) xu_c - G_m u_c. \tag{7.10}$$

Now finding

$$\frac{\partial e}{dx} = \frac{2s^2 - 36}{s^4 - 14s^2} u_c, \tag{7.11}$$

and taking the reference model to be

$$G_m(s) = \frac{1}{\frac{s^2}{\omega_n^2} + \frac{2\zeta}{\omega_n} s + 1}. \tag{7.12}$$

We want that the plant output y_p tracks y_m, output of the reference model, hence,

$$s^4 - 14s^2 \approx \frac{s^2}{\omega_n^2} + \frac{2\zeta}{\omega_n} s + 1, \tag{7.13}$$

which implies that

$$\frac{\partial e}{dx} = \left(\frac{1}{\frac{s^2}{\omega_n^2} + \frac{2\zeta}{\omega_n} s + 1}\right) u_c. \tag{7.14}$$

So, from the adaptation rule of (7.8) and (7.9), we get

$$\frac{dx}{dt} = -\gamma e \frac{de}{dx} = -\gamma \left(\left(\frac{1}{\frac{s^2}{\omega_n^2} + \frac{2\zeta}{\omega_n} s + 1}\right) u_c\right) e. \tag{7.15}$$

Choosing the value $\gamma = -0.01$ for integer control, $\gamma = -0.01$ for fractional controller and the reference model to be $G_M(s) = \frac{60}{s^2+15s+60}$ in designing the adaptive controller which can track the reference model in minimum time. The response of the Adaptive controller is shown in Figs. 7.2 and 7.3. From the figure it can be easily concluded that the adaptive fractional controller performs better in controlling the position of the cart, the angle of the pendulum and the controller response is tracking the desired response.

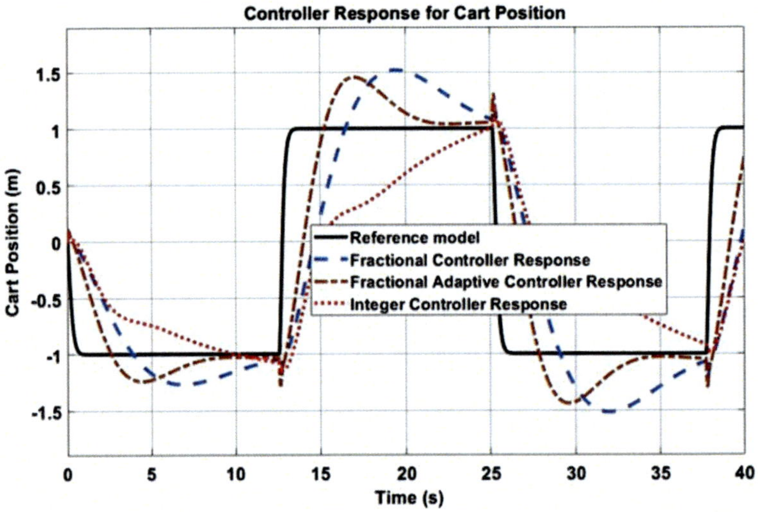

Fig. 7.2 Cart position adaptive control response

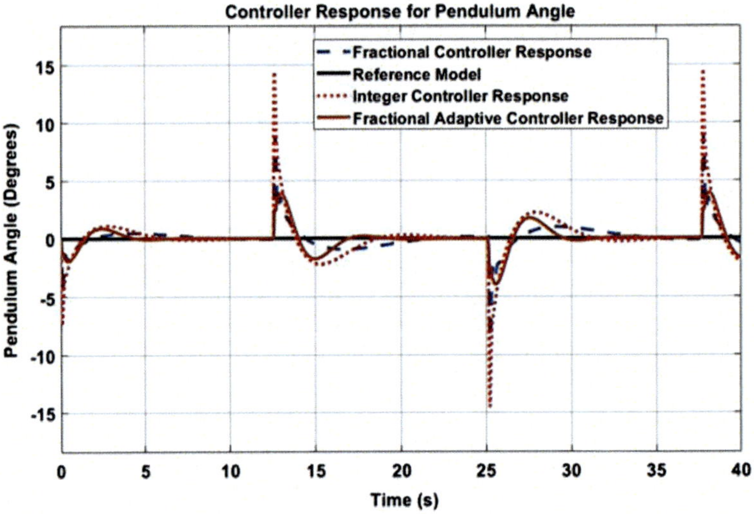

Fig. 7.3 Pendulum angle adaptive control response

7.3 Adaptive FO Controller Design for 2D Gantry Crane System

2D Gantry Crane System comes under under-actuated systems. It has one actuated input and two degrees of freedom. These cranes are used to pick heavy loads from one position and place to another position with zero or minimum swing. The main concern here is to control the position of the crane with a minimum swing.

Choosing the value $\gamma = 0.5$ for integer control, $\gamma = 0.5$ for fractional controller and the reference model to be $G_M(s) = \frac{60}{s^2+15s+60}$ in designing the adaptive controller which can track the reference model in minimum time. The response of the Adaptive controller is shown in Figs. 7.4 and 7.5. It can be easily concluded that the adaptive

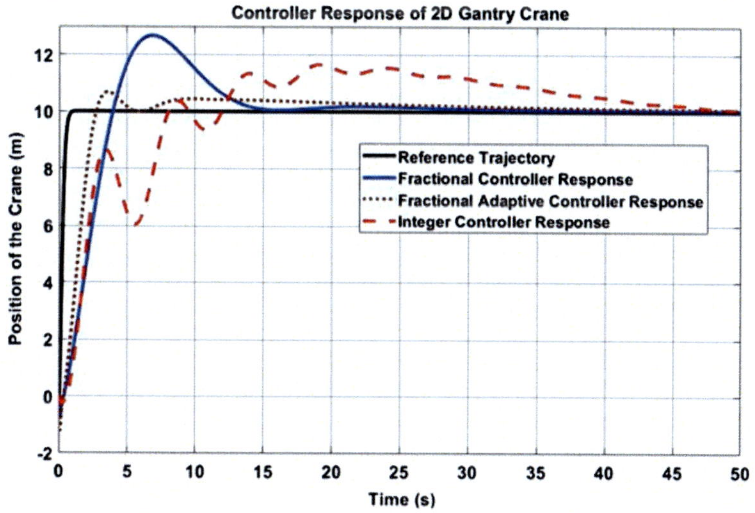

Fig. 7.4 Crane position adaptive control response

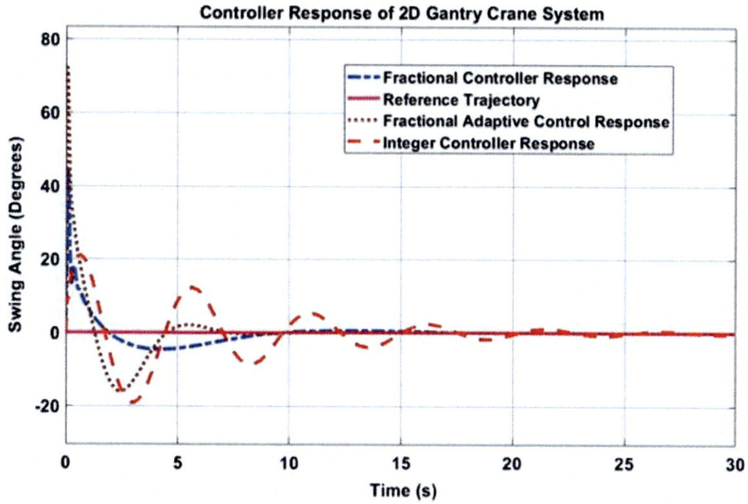

Fig. 7.5 Swing angle adaptive control response

fractional controller performs better in controlling the position of the crane, swing angle of the system and the controller response is tracking the desired response.

7.4 Adaptive FO Controller Design for Single Rigid Link Robotic Manipulator

A single rigid link robotic manipulators come under the fully actuated system. It has one actuator and one degree of freedom. A schematic diagram is shown in Fig. 7.6. Let us take the length of the link is l m, mass to be m kg and a constant gravity of g m/s^2. Taking frictional force B in consideration when a torque (input) 'τ' is applied. Then by Lagrangian (L), one can easily model this system [40, 41].

From Fig. 7.4 one can easily get the position of the arm tip,

$$\bar{r} = \begin{bmatrix} lcos\theta \ lsin\theta \end{bmatrix}, \ \dot{\bar{r}} = \begin{bmatrix} -lsin\theta\dot{\theta} \ lcos\theta \end{bmatrix}, \ \dot{\bar{r}}^2 = l^2\dot{\theta}^2. \tag{7.16}$$

Then the kinetic energy can be found to be

$$KE = \frac{1}{2} l^2 \dot{\theta}^2, \tag{7.17}$$

and potential energy

$$PE = mgl(1 - cos\theta). \tag{7.18}$$

and so

$$L = KE - PE = \frac{1}{2} l^2 \dot{\theta}^2 - mgl(1 - cos\theta), \tag{7.19}$$

and by EL formulation

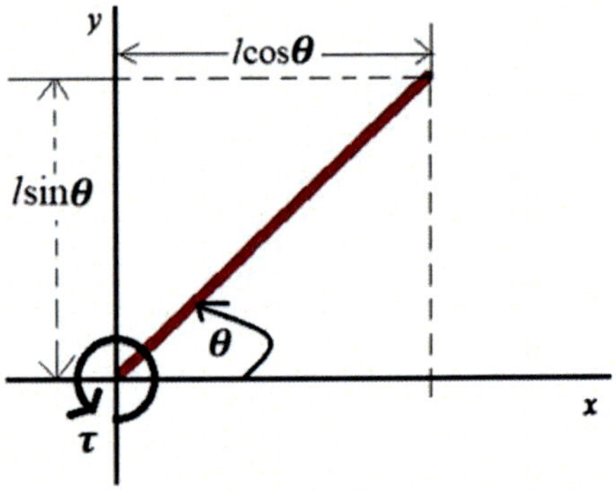

Fig. 7.6 Single rigid link robotic manipulator

$$\ddot{\theta} = -\frac{mg}{l} sin\theta - \frac{B}{l^2}\dot{\theta} + \tau. \tag{7.20}$$

Taking $l = 1m$, $m = 1\,\text{kg}$, $g = 10\,\text{m/s}^2$ and taking $B = 2$ and making them into state space equation form,

$$\dot{x_1} = x_2, \quad \dot{x_2} = -10\sin(x_1) - 2x_2 + \tau. \tag{7.21}$$

Then the linearised state space form will be

$$\begin{pmatrix}\dot{x_1}\\ \dot{x_2}\end{pmatrix} = \begin{pmatrix}0 & 1\\ -10 & -2\end{pmatrix}\begin{pmatrix}x_1\\ x_2\end{pmatrix} + \begin{pmatrix}0\\ 1\end{pmatrix}\tau, \tag{7.22}$$

with output equation as

$$y = (1\ 0)\begin{pmatrix}\theta\\ \dot{\theta}\end{pmatrix}. \tag{7.23}$$

Now, obtain the transfer function, and by using the same steps used from (7.6) to (7.22), an adaptive control response is easy to follow as shown in Fig. 7.7 which summarizes the adaptive controller design for single link robotic manipulator tracking the reference model efficiently when compared to the counterparts. Choosing the value $\gamma = -0.3$ for integer control, $\gamma = -0.3$ for fractional controller and the reference model to be $G_M(s) = \frac{60}{s^2+15s+60}$ in designing the adaptive controller which can track the reference model in minimum time. The response of the Adaptive controller is shown in Fig. 7.7. From the figure, it can be easily concluded that the

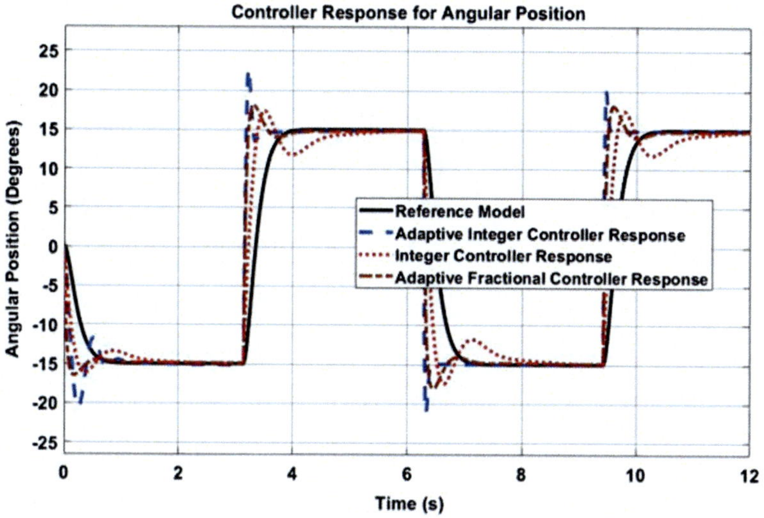

Fig. 7.7 Adaptive angle control response for single rigid link manipulator

adaptive fractional controller performs better in controlling the angular position of the manipulator and the controller response is tracking the desired response.

7.5 Adaptive FO Controller Design for 2DOF Serial Link Robotic Manipulator

2DOF serial link robotic manipulators come under fully-actuated systems. It has two actuated input and two degrees of freedom. These manipulators are used as robotic arms. The main concern here is to control the angular position of the robotic manipulator in minimum time.

Choosing the value $\gamma = 0.01$ for integer control of Link 1, the value $\gamma = 0.01$ for integer control of Link 2, $\gamma = 0.1$ for fractional controller of Link 1, $\gamma = 0.1$ for fractional controller of Link 2 and the reference model for both links to be $G_M(s) = \frac{60}{s^2+15s+60}$ in designing the adaptive controller which can track the reference model in minimum time. The response of the Adaptive controller is shown in Figs. 7.8 and 7.9. From the figure, it can be easily concluded that the adaptive fractional controller performs better in controlling the angular position of both the links and the controller response is tracking the desired response.

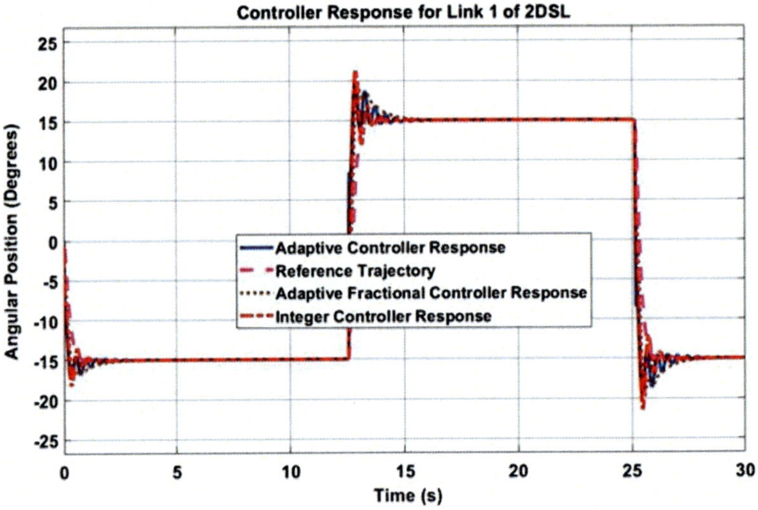

Fig. 7.8 Angular position adaptive control response for Link 1

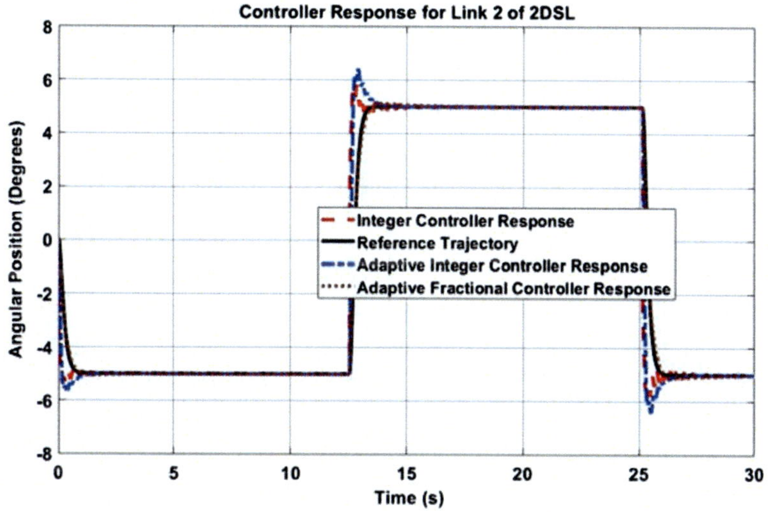

Fig. 7.9 Angular position adaptive control response for Link 2

7.6 Adaptive FO Controller Design for Missile Launching System (MLV)

MLV comes under fully-actuated systems. It has two actuated input and two degrees of freedom. The azimuth angle position, the MLV, and the firing angle decide the firing location. The idea here is to control the azimuth angle and firing angle base on the requirements in minimum time.

Choosing the value $\gamma = 0.001$ for integer control of azimuth angle, the value $\gamma = 0.02$ for integer control of firing angle, $\gamma = 0.001$ for fractional controller of azimuth angle, $\gamma = 0.1$ for fractional controller of firing angle and the reference model for both angles to be $G_M(s) = \frac{60}{s^2+15s+60}$ in designing the adaptive controller which can track the reference model in minimum time. The response of the Adaptive controller is shown in Figs. 7.10 and 7.11. From the figure, it can be easily concluded that the adaptive fractional controller performs better in controlling the azimuth and firing angles and the controller response smoothly tracks the desired response. The basic idea of how to design an adaptive fractional controller for robotic systems has been discussed. A Model Adaptive Reference Control (MARC) strategy is opted to find the control law. Pendulum on a cart system, 2D gantry crane system, Missile launching vehicle, single link robotic manipulator and 2 DOF serial link robotic manipulator has been considered to design the adaptive fractional controller design.

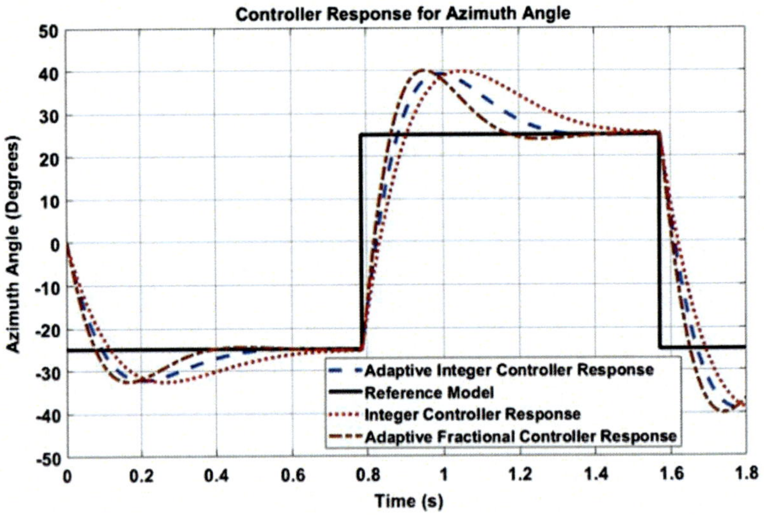

Fig. 7.10 Adaptive control response for azimuth angle of MLV

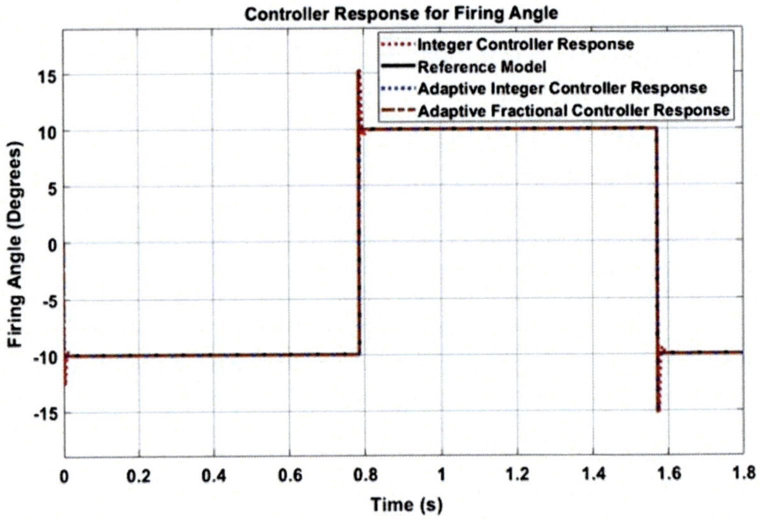

Fig. 7.11 Adaptive control response for firing angle of MLV

References

1. Patel, R., Deb, D., Dey, R., Balas, E., V., : Introduction to Adaptive Control. Intell. Syst. Ref. Libr. **53–65** (2019). https://doi.org/10.1007/978-3-030-18068-3_5
2. Nath, A., Deb, D., Dey, R., Das, S.: Blood glucose regulation in type 1 diabetic patients: an adaptive parametric compensation control-based approach. IET Syst. Biol. **12**(5), 219–225 (2018). https://doi.org/10.1049/iet-syb.2017.0093
3. Deb, D., Tao, G., Burkholder, J., Smith, D.: Adaptive compensation control of synthetic jet actuator arrays for airfoil virtual shaping. J. Aircr. **44**(2), 616–626 (2007). https://doi.org/10.2514/1.24910
4. Deb, D., Tao, G., Burkholder, J.: An adaptive inverse compensation scheme for signal-dependent actuator nonlinearities. In: 2007 46th IEEE Conference On Decision And Control (2007). https://doi.org/10.1109/cdc.2007.4434743
5. Liu, Y., Deb, D., Tao, G.: Modeling and multivariable adaptive control of aircraft with synthetic jet actuators. In: 2008 7th World Congress On Intelligent Control And Automation (2008). https://doi.org/10.1109/wcica.2008.4593263
6. Kapoor, D., Sodhi, P., Deb, D.: Adaptive failure compensation of a single main-rotor helicopter. IFAC Proc. **45**(1), 133–138 (2012). https://doi.org/10.3182/20120213-3-in-4034.00026
7. Patel, R., Deb, D., Modi, H., Shah, S.: Adaptive backstepping control scheme with integral action for quanser 2-dof helicopter. In: 2017 International Conference On Advances In Computing, Communications And Informatics (ICACCI) (2017). https://doi.org/10.1109/icacci.2017.8125901
8. Malepati, B., Vijay, D., Deb, D., Manjunath, K.: Parameter validation to ascertain voltage of Li-ion battery through adaptive control. Adv. Intell. Syst. Comput. **21–31**, (2018). https://doi.org/10.1007/978-981-13-1966-2_3
9. Deb, D.: Nonlinear adaptive control of vehicular radar servo systems with system and sensor uncertainties. In: 2011 IEEE Applied Electromagnetics Conference (AEMC) (2011). https://doi.org/10.1109/aemc.2011.6256910
10. Mehta, J., Dadi, Y., Patel, D., Shah, J., Deb, D.: In-vehicle adaptive traffic collision impact system for next-generation traffic. Innov. Res. Transp. Infrastruct. **77–87** (2018). https://doi.org/10.1007/978-981-13-2032-3_8
11. Rajkumar, S., Chakraborty, S., Dey, R., Deb, D.: Online delay estimation and adaptive compensation in wireless networked system: an embedded control design. Int. J. Control Autom. Syst. (2019). https://doi.org/10.1007/s12555-018-0612-x
12. Kapoor, D., Sodhi, P., Deb, D.: A novel control strategy to simulate solar panels. In: 2012 International Conference On Signal Processing And Communications (SPCOM) (2012). https://doi.org/10.1109/spcom.2012.6290002
13. Kapoor, D., Sodhi, P., Deb, D.: Solar panel simulation using adaptive control. In: 2012 IEEE International Conference On Control Applications (2012). https://doi.org/10.1109/cca.2012.6402674
14. Deb, D., Sonowal, S.: Synthetic jet actuator based adaptive neural network control of nonlinear fixed pitch wind turbine blades. In: 2013 IEEE International Conference On Control Applications (CCA) (2013). https://doi.org/10.1109/cca.2013.6662759
15. Patel, R., Deb, D., Dey, R., Balas, E.V.: Model reference adaptive control of microbial fuel cells. Intell. Syst. Ref. Libr. **109–121** (2019). https://doi.org/10.1007/978-3-030-18068-3_10
16. Patel, R., Deb, D., Dey, R., Balas, E.V.: Adaptive control of single chamber two-population MFC. Intell. Syst. Ref. Libr. **81–89** (2019). https://doi.org/10.1007/978-3-030-18068-3_7
17. Patel, R., Deb, D., Dey, R., Balas, E.V.: Adaptive and intelligent control of microbial fuel cells. In: Intelligent Systems Reference Library (2020). https://doi.org/10.1007/978-3-030-18068-3
18. Nath, A., Deb, D., Dey, R.: An augmented subcutaneous type 1 diabetic patient modelling and design of adaptive glucose control. J. Process Control **86**, 94–105 (2020). https://doi.org/10.1016/j.jprocont.2019.08.010

19. Nath, A., Deb, D., Dey, R.: Robust observer based adaptive control of blood glucose in diabetic patients. Int. J. Control **1–0** (2020). https://doi.org/10.1080/00207179.2020.1750705
20. Dhiman, H., Deb, D., Muresan, V., Balas, V.: Wake management in wind farms: an adaptive control approach. Energies **12**(7), 1247 (2019). https://doi.org/10.3390/en12071247
21. Deb, D., Tao, G., Burkholder, J., Smith, D.: An adaptive inverse control scheme for a synthetic jet actuator model. In: Proceedings Of The 2005, American Control Conference (2005). https://doi.org/10.1109/acc.2005.1470367
22. Deb, D., Tao, G., Burkholder, J., Smith, D.: Adaptive synthetic jet actuator compensation for a nonlinear tailless aircraft model at low angles of attack. In: 2006 American Control Conference (2006). https://doi.org/10.1109/acc.2006.1655376
23. Deb, D., Tao, G., Burkholder, J., Smith, D.: An adaptive inverse control scheme for synthetic jet actuator arrays. In: Infotech@Aerospace (2005). https://doi.org/10.2514/6.2005-7170
24. Deb, D., Tao, G., Burkholder, J.: High-order design of adaptive inverses for signal-dependent actuator nonlinearities. In: 2008 American Control Conference (2008). https://doi.org/10.1109/acc.2008.4587029
25. Deb, D., Tao, G., Burkholder, J., Smith, D.: Adaptive synthetic jet actuator compensation for a nonlinear aircraft model at low angles of attack. IEEE Trans. Control Syst. Technol. **16**(5), 983–995 (2008). https://doi.org/10.1109/tcst.2007.912124
26. Kapoor, D., Deb, D., Sahai, A., Bangar, H.: Adaptive failure compensation for coaxial rotor helicopter under propeller failure. In: 2012 American Control Conference (ACC) (2012). https://doi.org/10.1109/acc.2012.6315636
27. Patel, R., Deb, D.: Adaptive backstepping control of single chamber microbial fuel cell. IFAC-Papersonline **51**(1), 319–322 (2018). https://doi.org/10.1016/j.ifacol.2018.05.037
28. Patel, R., Deb, D., Dey, R., Balas, E.V.: Adaptive control of single population single chamber MFC. In: Adaptive and Intelligent Control of Microbial Fuel Cells. Intelligent Systems Reference Library, 161 (2020)
29. Patel, R., Deb, D.: Parametrized control-oriented mathematical model and adaptive backstepping control of a single chamber single population microbial fuel cell. J. Power Sources **396**, 599–604 (2018)
30. Patel, R., Deb, D.: Nonlinear adaptive control of microbial fuel cell with two species in a single chamber. J. Power Sources **434** (2019)
31. Wei, B.: Adaptive control design and stability analysis of robotic manipulators. Actuators **7**(4), 89 (2018). https://doi.org/10.3390/act7040089
32. He, W., Dong, Y., Sun, C.: Adaptive neural impedance control of a robotic manipulator with input saturation. IEEE Trans. Syst. Man Cybern. Syst. **46**(3), 334–344 (2016). https://doi.org/10.1109/tsmc.2015.2429555
33. Mondal, S., Mahanta, C.: Adaptive second order terminal sliding mode controller for robotic manipulators. J. Frankl. Inst. **351**(4), 2356–2377 (2014). https://doi.org/10.1016/j.jfranklin.2013.08.027
34. Yi, S., Zhai, J.: Adaptive second-order fast nonsingular terminal sliding mode control for robotic manipulators. ISA Trans. **90**, 41–51 (2019). https://doi.org/10.1016/j.isatra.2018.12.046
35. Nikdel, N., Badamchizadeh, M., Azimirad, V., Nazari, M.: Fractional-order adaptive backstepping control of robotic manipulators in the presence of model uncertainties and external disturbances. IEEE Trans. Ind. Electron. **63**(10), 6249–6256 (2016). https://doi.org/10.1109/tie.2016.2577624
36. Singh, A., Deb, D., Agarwal, H.: On selection of improved fractional model and control of different systems with experimental validation. Commun. Nonlinear Sci. Numer. Simul. **79**, 104902 (2019). https://doi.org/10.1016/j.cnsns.2019.104902
37. Ladaci, S., Charef, A., Loiseau, J.: Robust fractional adaptive control based on the strictly positive realness condition. Int. J. Appl. Math. Comput. Sci. **19**(1), 69–76 (2009)
38. Bouziane, K., Djouambi, A., Ladaci, S.: Fractional-order model reference adaptive controller design using a modified MIT rule and a feed-forward action for a DC-DC boost converter stabilization. In: 5th International Conference on Electrical Engineering, Boumerdes, pp. 1–6 (2017)

39. Ladaci, S., Charef, A.: On fractional adaptive control. Nonlinear Dyn. **43**(4), 365–378 (2006)
40. Shi, Z.X., Fung, E.H., Li, Y.C.: Dynamic modelling of a rigid-flexible manipulator for constrained motion task control. Appl. Math. Model. **23**(7), 509–525 (1999)
41. Green, A., Sasiadek, J.Z.: Dynamics and trajectory tracking control of a two-link robot manipulator. Modal Anal. **10**(10), 1415–1440 (2004)

Appendix A

A.1 Algorithm for Missile Launching Vehicle (MLV) System

A MATLAB code is given below to get linearised model of MLV system for different firing angle.

```
syms z1 z2 z3 z4 t1 t2 m1 m2 l1 l2 g
A=t1−m2∗l2∗l1∗(z4^2)∗(cos(z1)∗sin(z3)−sin(z1)
∗sin(z3));
B=t2+0.5∗(l2^2)∗m2∗(z4^2)∗sin(2∗z4)−m2∗l1∗l2∗(z2^2)
∗(cos(z1)∗cos(z3)+sin(z1)∗cos(z3))+m2∗g∗l2∗sin(z3);
G = 0.5∗(l2^2)∗m2∗(3+cos(2∗z3));
C = G∗(m1+m2)∗(l1^2)−(m2^2)∗(l1^2)∗(l2^2)
∗(sin(z1)∗cos(z3)−cos(z1)∗cos(z3))^2;
D = A;
E = B;
F = (m2^2)∗(l1^2)∗(l2^2)∗(sin(z1)∗cos(z3)−cos(z1)
∗cos(z3))^2 +(m1+m2)∗(l1^2)∗(1.5∗(l2^2)∗m2+0.5
∗(l2^2)∗m2∗cos(2∗z3));
f = [z2;((G∗A+m2∗l1∗l2∗(sin(z1)∗cos(z3)−cos(z1)
∗cos(z3))∗B)/C);z4;(m2∗l1∗l2∗(sin(z1)∗cos(z3)
−cos(z1)∗cos(z3))∗D−(m1+m2)∗(l1^2)∗E)/F];
v = [z1, z2, z3, z4];
R = jacobian(f, v)
b = jacobian(f, [t1 t2])
```


A. P. Singh et al., *Fractional Modeling and Controller Design of Robotic Manipulators*, Intelligent Systems Reference Library 194,
https://doi.org/10.1007/978-3-030-58247-0

```
_______________________________________________
%% Linear State Space

z1 = 0; z2 = 0; z3 = 0; z4 = 0; t1 = 0; t2 = 0; m1 =.2;

m2 = 20; l1=1; l2=2; g=9.8;
 H = R;
 J = b
 c = [1 0 0 0;0 0 1 0];
 d = [0,0;0,0]
_______________________________________________
```

A.2 Update Plant Model MATLAB Function

```
function [A,B,C,D,U,Y,X,DX] = fcn(m,u,x)

t = 0.1;

M=0.5;
l=1;
g=9.8;


Ac = [0 1 0 0;0 0 −m∗g/M 0;0 0 0 1;0 0 (M+m)∗g/(M∗l) 0];
Bc = [0 1/M 0 −1/(M∗l)]';
Cc = [1 0 0 0;0 0 1 0];
Dc = [0;0];


nx = size(Ac,1);
nu = size(Bc,2);
M = expm([[Ac Bc]∗Ts; zeros(nu,nx+nu)]);
A = M(1:nx,1:nx);
B = M(1:nx,nx+1:nx+nu);
C = Cc;
D = Dc;


X = x;
U = u;
Y = C∗x + D∗u;
DX = A∗x+B∗u−x;
```